Sugarcane - Its Products and Sustainability

Edited by Bimal Kumar Ghimire

Published in London, United Kingdom

Sugarcane - Its Products and Sustainability
http://dx.doi.org/10.5772/intechopen.100890
Edited by Bimal Kumar Ghimire

Contributors
Evrim Özkale, Linga Banoth, Suresh Babu Bastipati, Khawlhring Rosangzuala, Patlolla Ravinder Reddy, Cheemalamarri Chandrasekhar, Raveena Gajjala, Patapanchala Ravindra Babu, Ramesh Malothu, Prakasham Reddy Shetty, Ravichandra Korasapati, Daniel Henrique Dario Capitani, Fernando Chiñas-Castillo, Rafael Alavez-Ramirez, Magdaleno Caballero-Caballero, Valentín Juventino Morales-Dominguez, Margarito Ortiz-Guzman, Roberto Candido Jimenez-Piñon, Maria Eugenia Silva-Rivera, Angel Ramos-Alonso, Muhammad Ehsan Khan, Shahid Afghan, Waqas Raza Arshad, Karim Bukhsh Malik, Amin Nikpay, Mounia Elhaddadi, Abhisek Shrestha, Bharti Thapa, Ganga Dulal

First published in London, United Kingdom, 2023 by IntechOpen
IntechOpen is the global imprint of INTECHOPEN LIMITED, registered in England and Wales, registration number: 11086078, 5 Princes Gate Court, London, SW7 2QJ, United Kingdom

British Library Cataloguing-in-Publication Data
A catalogue record for this book is available from the British Library

Additional hard and PDF copies can be obtained from orders@intechopen.com

Sugarcane - Its Products and Sustainability
Edited by Bimal Kumar Ghimire
p. cm.
Print ISBN 978-1-80356-368-8
Online ISBN 978-1-80356-369-5
eBook (PDF) ISBN 978-1-80356-370-1

Meet the editor

Dr. Bimal Kumar Ghimire is currently an assistant professor in the Department of Crop Science, College of Sanghuh Life Science, Konkuk University, Seoul, South Korea. He received his MS in Botany (major in Cytogenetics) from North Bengal University, India. He obtained his Ph.D. in Agriculture Science from Kangwon National University, South Korea, in 2008. Subsequently, he joined the Department of Botany, Sikkim University, India, as an assistant professor. He has published more than 100 articles in peer-reviewed journals and books. He has edited and written several books and contributed chapters on different aspects of plant sciences including light sources generally used in controlled agriculture, genetic transformation, and secondary metabolites of medicinal plants. His research interests and experience include plant anatomy, genetic diversity in bioenergy crops, phytochemical compounds, tissues and cell culture, bioreactor culture, and genetic transformation of important medicinal plants.

Contents

Preface

Sugarcane (*Saccharum officinarum* L.) is one of the main bioenergy crops and agricultural commodities worldwide. Sugarcane biomass is a promising source of sustainable renewable energy to meet the growing demands of biofuel. This book discusses the bioremediation properties of wastewater sugar refineries. It presents comprehensive information about current advances in value-added products of sugar cane bagasse. As such, it also provides updates on the challenges and potentiality of sugar-ethanol and second-generation ethanol-electricity cogeneration. It contains seven chapters by eminent research scholars that cover a vast range of research on sugarcane as a construction material, its economic importance, and its yield potential. The book also highlights the social role of sugarcane cultivation. Finally, the book discusses the importance of abiotic stress affecting sugarcane productivity and its morphological characteristics. We are enormously grateful to our publisher IntechOpen and the authors who contributed their work to this book. We also extend our sincerest appreciation to Editor and Author Service Manager Ms. Romina Rovan who was instrumental in the preparation and publication of this volume.

Dr. Bimal Kumar Ghimire
Department of Crop Science,
College of Sanghuh Life Science,
Konkuk University,
Seoul, South Korea

Section 1

Sugarcane Biorefinery and Bioremediation of Wastes

Chapter 1

Bioremediation of Wastewaters of Sugarcane Biorefineries

Evrim Özkale

Abstract

In addition to sugar, sugarcane biorefineries emerge as the integration of different sugarcane industries and produce several wastes and wastewaters that are disposed into the environment. In biorefineries and other facilities, these wastes are used to produce several chemicals, including bioplastics and bioethanol. However, these industries use a greater amount of fresh water and their effluents, which have high amounts of solids and are released mostly into water or used for crop irrigation. Inadequately treated and indiscriminately disposing and discharging of effluents to the environment contributes to a greater risk of pollution of soils and waters. Therefore, to minimize the side effects, control the environmental burden, grow the beneficiaries of waste, and sustain a healthy environment for the future, proper management of industrial wastes is important.

Keywords: sugar industry, industrial wastewater, wastewater treatment, biological treatment efforts

1. Introduction

The sugar industry is the most common industry in more than 130 countries, especially in developing countries [1, 2]. Sugarcane processing plants comprise one of the most successful examples of biorefineries in which a wide range of products is obtained from a single raw material of feedstock. Biorefinery is emerging as the integration of the different sugarcane industries based on biomass feedstock.

Harvested sugarcane is used to produce raw sugar in addition to products, such as bagasse (residue from sugarcane crushing), press mud (dirt mud from juice clarification), molasses (final residue from sugar crystallization), and wastewater [3]. Production of ethanol concomitantly is in a greater ratio than to produce only raw sugar or ethanol in many countries around the world [2].

Biorefinery has emerged as the integration of different sugarcane industries based on biomass feedstock. However, ethanol distilleries are growing at an alarming rate across the globe and generate a huge volume of effluent that is disposed/discharged into the environment. These inadequately/non-treated and indiscriminately disposed effluents cause pollution of soil and water resources or deposits [4].

The agricultural industries are the major freshwater consumers and play a major role in polluting water bodies. Freshwater is used in different units of sugar production processes and generates wastewater, which is highly variable both in quantity and quality. Regarding the processes of the feedstocks, products, and chemicals used in the process, wastewaters/effluents have different characteristics [5, 6].

Different pollution monitoring agencies, such as State and National Pollution Control Boards, have been made compulsory for each industry to set up wastewater treatment plants. In the treatment system, simple treatments of effluent are not effective to the dischargeable limit.

Sugarcane processing plants comprise one of the most successful examples of biorefineries in which a wide range of products is obtained from a single raw material of feedstock. Recently, bioplastic production has emerged as one of the primary interests in the sugar processing industry in which the sugars are converted into lactic acid and polymerized into biopolymer [7].

Wastewaters are treated by several methods, such as adsorbent, electrochemical, anaerobic biological treatment, and biochemical oxidation. However, treated wastewater by these methods are not meeting the discharge limit, therefore requires modification. Electrolysis followed by coagulation is the most effective method of treatment of sugar industry wastewater giving close to 100% reduction of COD as well as the electrochemical process shows 81% COD and 83.5% color reduction at pH 6.0, electrode distance of 20 mm. The combined treatment results show 98% COD and 99.5% color removal at 8 mM mass loading and pH 6 with copper sulfate [8, 9].

The preferred choice for the treatment of effluents is anaerobic degradation because these industries typically generate high-strength wastewater with the potential to recover energy in the form of biogas [10]. Anaerobic treatment processes have been employed to stabilize sewage sludge for more than a century. The application of this process for high-strength industrial wastewater treatment began with the development of high-rate anaerobic reactors (HR). Up-flow anaerobic sludge blanket (UASB) reactor is used in treating sugar wastewater with varying HRT and varying feed concentrations at ambient conditions [11].

AD-based biorefineries have great potential to meet the energy deficiency criteria of sustainable development goals in the coming decades. Following the need to increase energy efficiency and diversify the product portfolio derived from sugarcane, anaerobic digestion (AD) may be an old solution with great potential to improve the biorefinery character of the sugarcane industry. Therefore, sugarcane biorefineries and wastewater treatments are required for proper management and for environmental sustainability [1, 12].

2. Characteristics of wastewaters of sugar industry and effluents of ethanol distilleries

The sugar industry is one of the major industries, which has been included in the polluting industries list by the World Bank [13]. In the sugar industry, wastewater is produced from sources primarily due to cane processing during evaporation, crystallization, and refinery, and the other is from cycling, such as in the condensers, chimneys, scrubbers, and refrigeration of turbines [14]. To crush one ton of sugarcane, nearly 2000 l of water is required, which generated nearly 1000 l of wastewater. Moreover, per ton of crushed sugarcane, 0.7 m^3 of wastewater is produced due to the high moisture content of the raw material [15, 16]. Also, the wastewater for periodically cleaning procedures, such as cleaning of lime water and SO_2, producing house of the industry is contributed.

A large amount of wastewater as effluent, which has a high amount of solids, BOD_5, COD, chloride, sulfate, nitrate, and magnesium, was also discharged into the lands beside water [17]. The major sources of wastewater in molasses-based distilleries are fermenter sludge, spent wash, and spent less. The fermented wash is the main product of fermentation which is decanted; the remaining sludge is known as yeast sludge (fermenter sludge) and contributes to the pollution load from the distilleries. On the other hand, spent wash is a complex effluent having the strongest organic matter in terms of having high COD (up to 160,000 mg/L), temperature, low acidity (between 3.7 and 4.5), and a high content of dissolved inorganic salts and ash content [4, 18].

Principally, the biological treatment method is effective for highly polluted agro-industrial wastewater from the sugar industries and ethanol distilleries [19]. However, the current anaerobic treatment technologies and the high-rate anaerobic reactors are the most suitable and attractive primary treatment options for high-strength organic effluents, such as sugar industry wastewater and distillery spent wash [20].

More than 25 tons COD of agro-industrial waste (water), on a daily basis, can be converted into 7000 m^3CH_4 (80% CH_4 recovery approximately) that accounts for an

Parameter	Sugar Industry	Effluent of ethanol distilleries spent wash	Anaerobically treated effluent quality
Temperature (°C)	29.3–44.3	46.3–56.3	
pH	6.7–8.4	3.9–4.9	7.5–8.0
Electric conductivity	540.3–925.9	3910.0–50,500.00	
BOD_5	654.6–1968.5	50,000.00–60,000.00	8000–10.000
COD	1100.3–2148.9	110,000.00–190,000.00	45,000–52,000
Chloride	30.5–866.6	6213.6–7475.7	
Total hardness	356.2–2493.1	3100.3–4477.2	
Calcium	365.4–468.0	8000–8500	
Magnesium	214.8–341.0	816.3–1828.1	
Total solids	2452.3–3050.6	91,876.9–150,300.9	70,000–75,000
Total dissolved solids	1480,2–1915.1	13.000.0–88,265.00	30.000–32.000
Total suspended solids	220.3–790.7	3611.1–150,000.00	
Nitrates	0.4–0.9	2.40–32.9	
Organic- N	24.3–36.4	75.2–400.7	
Ammonia- N	0.0–4,.2	10.9–18.1	
Total Nitrogen	11.1–40.6	85.8–1355.3	
Phosphate	1.2–9.6	2500–2700	1500–1700
Sulfate	21.5–51.7	803–6050.5	
Oil and greases	**88.7–134.4**	**3.3–202.1**	

Table 1.
Characteristics of wastewater of sugar industry and effluents of ethanol distilleries and some quality parameters of anaerobically treated effluent (mg/L) [1].

energy equivalent of about 250 GJ/d working with a modern combined heat power (CHP) gas engine, due to 40% of efficiency, a useful 1–2 MW electric output can be achieved. By this conversion, CO_2 emission reduction (ton $CO_2/m^3.y$), is based on coal-driven powerplant (**Table 1**) [21, 22].

3. Wastewater management and biological treatment

The most common ways of wastewater management in the sugar industry and biorefineries are fertirrigation, bio-compost, and concentration by evaporation (incineration). However, these conversions are difficult to manage since the huge volumes of wastes produced during the sugar production and biorefinery processes are costly. Also, prior to producing bio-compost, BOD_5 and COD of the distillery spent wash should be anaerobically digested (biomethanation). As the industrial waste is converted into organic-rich manure, not only the problems of waste disposal and pollution are solved but also the soils are replenished and renovated. However, raw distillery spent wash has to be subjected primarily to anaerobic digestion (biomethanation) treatment to decrease BOD_5 and COD and other pollutants before combining with the press mud to produce bio-compost (**Figure 1**) [23, 24].

There is a growing interest in biological treatment systems as a common procedure for wastewater treatment to eliminate solids, nutrients, and organic matter since the various conventional physicochemical methods have been tested and are found as inefficient and had some drawbacks for the treatment of sugar industry wastewaters.

In the usual procedure, biological wastewater treatment processes use microorganisms to utilize wastewater pollutants for their growth and metabolize the organic substrate in the wastewater to gain their energy and into metabolic wastes, such as CO_2 and water [25]. Biological wastewater treatment processes are beneficiary alternatives

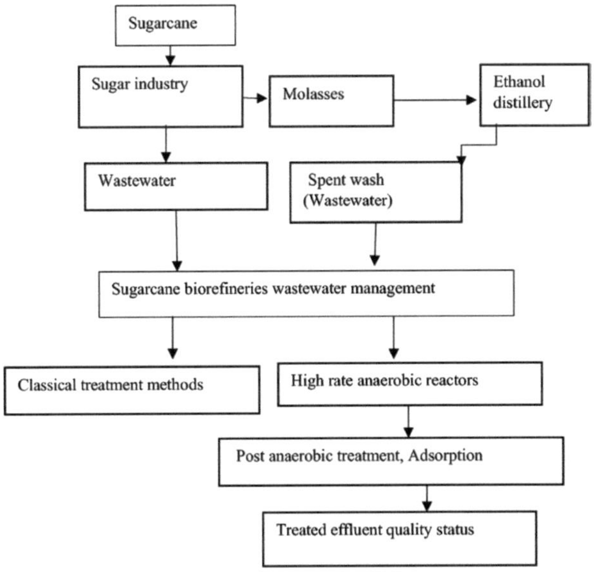

Figure 1.
Sugar industry wastes and wastewater management and biological treatment [1].

to conventional treatments. Methods, such as anaerobic bioremediation, employed to remove various pollutants from several industrial wastewaters can be applicable to wastewaters of sugar industries and ethanol distilleries spent wash. Additionally, the removal of pollutants and toxic materials from industrial wastewater is increasingly shifting from the use of conventional approaches to the implementation of the advanced bioremediation processes of high-rate anaerobic conditions [26].

In recent years, fermentation reactors have been the preferred method for treating sugar industry effluent, encompassing technologies, such as anaerobic fixed bed (UAFB), up-flow anaerobic sludge blanket (UASB), anaerobic downflow stationary fixed film (DSFF), aerated fixed film (AFF), and anaerobic batch reactors [27]. High-rate anaerobic reactors are the most suitable treatment option for high-strength organic effluents and have characteristics, such as high loading capacity, better stability, resistance to inhibitors, and low sludge production. One disadvantage of the fermentation methods is that the anaerobic process is characterized by low nitrogen and phosphorus removal rates. The fluidized active filling (FAF) method was presented as an innovative solution to treat sugar industry effluent in anaerobic reactors. It was shown that the use of the FAF-reactor could improve phosphorus removal rates [27].

The molasses distillery wastewater has similarities with sugar industry wastewater in composition, but the distillery spent wash is high-strength wastewater. Recent studies have proven the suitability of anaerobically processing sugarcane vinasse on technical, economic, and environmental bases. The exploitation of AD plants within the sugarcane biorefinery has provided reliable data to encourage energetic exploitation and economic feasibility of sugarcane vinasse.

There is a continuous search to combine clean energy supply with the exploitation of bioenergy sources to meet global energy demands on an environmentally friendly basis. On this basis, sugarcane biorefineries have a central role by providing sugar, ethanol, and bioelectricity [28].

The extensively used technology—UASB technology—is efficient and commonly used for also the sugar industry effluent and distillery spent wash. UASB is effective among the high-rate anaerobic digestion methods, but its limitations are found, such as longer HRTs, long periods of startup, and the wash-up of the sludge, in the bioreactor [17, 29].

Even though this treatment technology is very promising for the wastewater treatment sector, it was observed that its efficiency is somehow lower as compared to conventional biological treatment systems [30]. Therefore, the application of this treatment option is not feasible enough for the high strength of the sugar industry and ethanol distillery wastewater, yet. The maximum BOD_5 reduction (86%) occurred reported at a BOD_5 loading rate of 2.74 kg/m^3 and a digestion temperature of 50°C [17].

The energy returned on energy invested (EROEI) ratio associated with sugarcane-derived 1G ethanol is often reported to exceed a value of 6.5. The effective EROEI ratio obtained for ethanol production should be lower than 2.0 without considering the contribution of bagasse burning [31].

Applications apart from electricity generation may also comprise energetically efficient alternatives to exploit biogas focusing on biomethane ($bioCH_4$) production. $BioCH_4$ in which the maximum methane content is 96.5% (v/v), may be used *in loco* diesel replacement and/or injected into the gas grid. Previous studies indicated that $bioCH_4$ production is the most attractive approach (compared to electricity production) on economic bases for vinasse-fed AD plants [32].

Using the anaerobic processing of vinasse tends to increase the EROEI in ethanol-producing plants by less than 20.0%, whilst investing in biogas production from raw feed-stocks leads to an approximately 400% increase in the sugarcane biorefinery. Different designs for AD plants have been applied to vinasse in large-scale distilleries for enhancement of energy recovery. The application of chemicals to raw vinasse is a key factor for obtaining high treatment performances and specific composition characteristics, such as high carbohydrate concentrations, low pH, and absence of alkalinity provide favorable conditions for rapid vinasse acidification. Sodium bicarbonate has been frequently applied as an alkalinizing component in bench-scale AD systems for vinasse treatments [32–34]. Sodium hydroxide and calcium carbonate have also been commonly used in treating vinasse prior to methane production. However, the use of high chemical doses negatively affects the economic favorability of scaling up AD plants. Urea application has also been tested to support alkalinity to methanogens and to increase nitrogen levels in the treated vinasse thereby increasing its fertilization potential. However, the potential of accumulation of ammonia within the reactors could lead to inhibitory effects over methanogenic populations, thus impairing the bioenergy recovery from vinasse [35].

The combined use of low dosages of chemicals for the recirculation of the effluent may offer the most suitable alternative for the alkalinization of full-scale AD plants on an economic and environmental basis. Further studies on alkalizing strategies for vinasse are under consideration as an approach to improve both the economic and environmental performances of biodigestion plants [32, 36].

On the other hand, AD alone cannot remove all the pollutants of sugarcane biorefinery wastewater. As post-treatment, another treatment technology is highly recommended for the removal of less biologically degradable organic compounds in terms of the treatment efficiency, cost, and ease of operation as well as social acceptance. As a potential candidate for post-anaerobic treatment of sugarcane biorefinery wastewater, adsorption treatment technology is considered mostly [1].

In recent years, the integration of microalgal cultivation with other processes for achieving inexpensive nutrient and energy use has become an important issue. Integrating mixotrophic microalgae into wastewater treatment is a cost-effective and feasible method for biofixation of CO_2 [37, 38]. The principal advantage of incorporating microalgae into wastewater treatment is the generation of O_2 through photosynthesis and is necessary for heterotrophic bacteria to biodegrade carbonaceous material. The potential of the fuel gas and the wastewater of a sugar factory to support microalgae growth for biofuel and biofertilizer production has also been evaluated. In addition to removing pollutants, the cultivation of microalgae in conjunction with wastewater treatment provided lipids that can be converted into biodiesel.

4. Conclusions

Sugarcane processing industries are highly water-intensive processes. Proper water conservations and management of cane industrial wastewater are highly important for sustainable freshwater uses.

Conventional treatment methods are challenging in the elimination of pollutants. Recently high-rate anaerobic digestion has been recognized as a safe and effective treatment, particularly for wastewaters of the organic saturated sugar industry and ethanol distilleries. Although economic and environmental assessments are still required to fully subsidize the demonstration of the advantages of the AD-based biorefinery, the results of the studies encourage more efficient renewable systems. The proper management of vinasse is standing as one of the major issues facing sugarcane-based ethanol biorefineries due to environmental concerns.

The anaerobic treatment of wastewater is an important field of research where improvements and new developments are needed to overcome the limitations of the system. Technological solutions allowing the increase of the technological efficiency of anaerobic methods of wastewater treatment are still under investigation.

Author details

Evrim Özkale
Manisa Celal Bayar University, Manisa, Turkey

*Address all correspondence to: evrimmiko@gmail.com

References

[1] Fito J, Tefera N, van Hulle SWH. Sugarcane biorefineries wastewater: Bioremediation technologies for environmental sustainability. Chemical and Biological Technologies in Agriculture. 2019;**6**:6. DOI: 10.1186/s40538-019-0144-5

[2] Fito J, Tefera N, Demeku S, Kloos H. Water footprint as an emerging environmental tool for assessing sustainable water use of the bioethanol distillery at Metahara sugarcane farm, Oromiya region, Ethiopia. Water Conservation Science Engineering. 2017;**2**:165-176

[3] Nandy T, Shastry S, Kaul SN. Wastewater management in a cane molasses distillery involving bioresource recovery. Journal of Environmental Management. 2002;**65**:25-38

[4] Fito J, Tefera N, Van Hulle SWH. Physicochemical properties of the sugar industry and ethanol distillery wastewater and their impact on the environment. Sugar Technology. 2019;**21**:265-267. DOI: 10.007/s12355-018-0622-z

[5] Agarwal MK, Kumar M, Ghosh P, Kumar SS, Sigh L, Vijay VK, et al. Anaerobic digestion of sugarcane bagasse for biogas production and digestate valorization. Chemosphere. 2022;**295**:133893. DOI: 10.1016/j.chemosphere.2022.133893

[6] Klein BC, Chagas MF, Watanabe MDB, Bonomi A, Filho RM. Low carbon biofuels and the new Brazilian National Biofuel Policy (RenovaBio): A case study for sugarcane mills and integrated sugarcane-microalgae biorefineries. Renewable and Sustainable Energy Reviews. 2019;**115**:109365. DOI: 10.1016/j.rser.2019.109365

[7] Sriroth K, Vanichsriratana W, Sunthornvarabhas J. The current status of sugar industry and by-products in Thailand. Sugar Technology. 2016;**18**:576-582

[8] Fuess LT, Klein BC, Chagas MF, Rezende MCAF, Garcia ML, Bonomi A, et al. Diversifying the technological strategies for recovering bioenergy from the two-phase anaerobic digestion of sugarcane vinasse: An integrated techno-economic and environmental approach. Renewable Energy. 2018;**122**:674-687

[9] Konde KS, Nagarajan S, Kumar V, Patil SV, Ranade E. Sugarcane bagasse based biorefineries in India: Potential and challenges. Sustainable Energy and Fuels. 2021;**5**:52-78. DOI: 10.1039/d0se01332c

[10] Breitenmoser L, Gross T, Huesch R, Rau J, Dhar H, Kumar S, et al. Anaerobic digestion of biowastes in India: Opportunities, challenges and research needs. Journal of Environmental Management. 2019;**236**:396-412

[11] Surendra KC, Sawatdeenarunat C, Shresta S, Sung S, Kumar SK. Anaerobic digestion-based biorefinery for bioenergy and biobased products. Industrial Biotechnology. 2015;**11**:2. DOI: 10.1089/ind.2015.0001

[12] Liu Z, Wei L, Liu Y. A sustainable biorefinery to convert agricultural residues into value-added chemicals. Biotechnology for Biofuels and Bioproducts. 2016;**9**:197. DOI: 10.1186/s13068-016-0609-8

[13] Tiwari A, Sahu O. Treatment of food-agro (sugar) industry wastewater

with copper metal and salt: Chemical oxidation and electro-oxidation combined study in batch mode. Water Resources and Industry. 2017;**17**:19-25. DOI: 10.1016/j.wri.2016.12.001

[14] Ingaramo A, Heluane H, Colombo M, Cesca M. Water and wastewater eco-efciency indicators for the sugar cane industry. Journal of Cleaner Production. 2009;**17**:487-495

[15] Memon AR, Suhail Ahmed S, Abdul Khaliq A. Sugar industry efuent characteristics and chemical analysis. Journal of Applied Emerging Science. 2006;**1**:156-157

[16] Solomon SK. Environmental pollution and its management in sugar industry in India: An appraisal. Sugar Technology. 2005;**7**:77-81

[17] Mohana S, Acharya BK, Madamwar D. Distillery spent wash: Treatment technologies and potential applications. Journal of Hazardous Materials. 2009;**163**:12-25

[18] Basu S, Mukherjee S, Kaushik A, Batra VS. Integrated treatment of molasses distillery wastewater using micro filtration (MF). Journal of Environmental Management. 2015;**158**:55-60

[19] Pant D, Adholeya A. Biological approaches for treatment of distillery wastewater: A review. Bioresource Technology. 2007;**98**:2321-2334

[20] Rajeshwari KV, Balakrishnan M, Kansal A, Lata K, Kishore VVN. State-of the-art of anaerobic digestion technology for industrial wastewater treatment. Renewable and Sustainable Energy Reviews. 2000;**4**:135-156

[21] Fuess LT, Zaiat M. Economics of anaerobic digestion for processing sugarcane vinasse: Applying sensitivity analysis to increase process profitability in diversified biogas applications. Process Safe Environment Protection. 2018;**115**:27-37

[22] Fuess LT, Kiyuna LSM, Ferraz ADN Jr, Persinoti GF, Squina FM, Garcia ML, et al. Thermophilic two-phase anaerobic digestion using an innovative fixed-bed reactor for enhanced organic matter removal and bioenergy recovery from sugarcane vinasse. Applied Energy. 2017;**189**:480-491. DOI: 10.1016/j.apenergy.2016.12.071

[23] Ghulam S, Khan MJ, Usman K. Effect of different rates of pressmud on plant growth and yield of lentil in calcareous soil. Sarhad Journal of Agriculture. 2012;**28**:8-11

[24] Analia A, Juliana MS, Costa SD, Colin VL, Fuentes MS, Sergio Antonio C, et al. Actinobacteria: Current research and perspectives for bioremediation of pesticides and heavy metals. Chemosphere. 2017;**166**:41-62

[25] Kharayat Y. Distillery wastewater: Bioremediation approaches. Journal of Integrative Environmental Sciences. 2012;**9**:69-91

[26] Igwe JC, Abia AA. A bioseparation process for removing heavy metals from waste water using biosorbents. African Journal of Biotechnology. 2006;**5**:1167-1179

[27] Debowski M, Zieli'nski M. Technological effectiveness of sugar industry effluent methane fermentation in a fluidized active filling reactor (FAF-R). Energies. 2020;**13**:6626. DOI: 10.3390/en13246626

[28] Fuess LT, Cruz RBCM, Zaiat M, Nascimento CAO. Diversifying the portfolio of sugarcane biorefineries:

Anaerobic digestion as the core process for enhanced resource recovery. Renewable and Sustainable Energy Reviews. 2021;**147**:111246. DOI: 10.1016/j.rser.2021.111246

[29] Acharya BK, Mohana S, Madamwar D. Anaerobic treatment of distillery spent wash—A study on upfow anaerobic fxed flm bioreactor. Bioresource Technology. 2008;**99**:4621-4626

[30] Yasar A, Tabinda AB. Anaerobic treatment of industrial wastewater by UASB reactor integrated with chemical oxidation processes; an overview. Polish Journal of Environmental Studies. 2010;**19**:1051-1061

[31] Chiriboga G, de la Rosa A, Molina C, Velarde S, Carvajal CG. Energy return on investment (EROI) and life cycle analysis (LCA) of biofuels in Ecuador. Heliyon. 2020;**6**:e04213. DOI: 10.1016/j.heliyon.2020.e04213

[32] Fuess LT, de Arajuo Junior MM, Garcia ML, Zaiat M. Designing full-scale biodigestion plants for the treatment of vinasse in sugarcane biorefineries: How phase separation and alkalinization impact biogas and electricity production costs? Chemical Engineering Research and Design. 2017;**119**:209-220

[33] Ferraz ADN Jr, Koyama MH, Araújo MM Jr, Zaiat M. Thermophilic anaerobic digestion of raw sugarcane vinasse. Renewable Energy. 2016;**89**:245-252

[34] Siqueira LM, Damiano ESG, Silva EL. Influence of organic loading rate on the anaerobic treatment of sugarcane vinasse and biogas production in fluidized bed reactor. Journal of Environmental Science and Health, Part A. 2013;**48**(13):1707-1716. DOI: 10.1080/10934529.2013.815535

[35] Bonc MA, Formagini E, Santos LS, Marques RD, Paulo PL. Application of urea dosing for alkalinity supply during anaerobic digestion of vinasse. Water Science and Technology. 2012;**66**(11):2453-2460. DOI: 10.2166/wst.2012.476

[36] van Haandel AC. Integrated energy production and reduction of the environmental impact at alcohol distillery plants. Water Science and Technology. 2005;**52**(1-2):49-57

[37] Chen Y, Ho S, Nagarajan D, Ren N, Chang J. Waste biorefineries- integrating anaerobic digestion and microalgae cultivation for bioenergy production. Current Opinion in Biotechnology. 2018;**50**:101-110

[38] Silva MA, Barbosa GH, Codato CB, Mattos LFA, Bastos RG, Kieckbusch TG. Heterotrophic growth of green microalgae *Desmodesmus subspicatus* in ethanol distillation wastewater (vinasse) and lipid extraction with supercritical CO_2. Journal of Chemical Technology and Biotechnology. 2017;**92**:573-579

Chapter 2

Value Added Products Generation from Sugarcane Bagasse and Its Impact on Economizing Biorefinery and Sustainability of Sugarcane Industry

Ravichandra Korasapati, Suresh Babu Basthipati, Khawlhring Rosangzuala, Patlolla Ravinder Reddy, Raveena Gajjala, Patapanchala Ravindra Babu, Cheemalamarri Chandrasekhar, Ramesh Malothu, Prakasham Reddy Shetty and Linga Banoth

Abstract

Augmenting value-added products generation with the biorefinery process of sugar cane by utilizing the by-products helps to achieve a more sustainable model of the sugarcane industry and in turn, contributes to the circular economy. Among the value-added products produced from sugarcane waste, functional foods offer additional health benefits besides their nutritional and calorific value. In recent years non-digestible sugars gained interest as potential prebiotic functional foods which benefit the host without increasing calorific value. These sugars are produced by the breakdown of carbohydrate polymers like cellulose and xylan, by thermochemical treatment or by enzymatic hydrolysis, or a combination of both. Sugar cane bagasse (SB) is an economical source of xylan which can serve as the substrate for xylooligosaccharides (XOS), xylobiose, xylitol, and ethanol. Cellulases, xylanases, and ligninases have wide applications in food processing, agro-fiber, pharmaceutical, and the paper and pulp industries including nutraceuticals production, where these enzymes provide eco-friendly alternatives to some chemical processes and help to reduce environmental impact. Conventional thermochemical methods for nutraceuticals production require chemicals that result in the release of toxic byproducts thus requiring additional steps for refining. In this context, the sustainable and eco-friendly processes for the production of nutraceuticals require employing biocatalysts like microbial enzymes or microbes as a whole, where in addition to averting the toxic byproducts the refining process requires lesser steps. The present chapter discusses the current research and challenges in the production of value-added products from sugarcane byproducts and their contribution to the sustainability of the sugarcane industry.

Keywords: sugar cane industry, bagasse, value added products, sustainable processing

1. Introduction

The quest for environmentally friendly and economically feasible alternatives for conventional fuel resources was necessitated by the increase in price hikes and climate change caused by fossil fuels for many years [1]. An increase in the production cost of ethanol due to an increase in prices of food/feed-derived substrates (food grains, sugar molasses) poses a challenge to sustainability in a long run. These challenges have forced researchers to look into other approaches for a biofuel supply that is both affordable and sustainable. Sugarcane straw, sugarcane pulp/bagasse, and sugarcane residues may be great options of feedstock for the production of second-generation (2G) biofuel (**Figure 1**) [2].

The significant productivity gains, product diversification, and a decrease in the environmental effect of fields of sugarcane crop, sugar industries, distillation facilities, and traditional sugar manufacturing by evaporation, resulted in the global sugar sector currently exhibiting inertia in production and weak sustainability [6]. The inclusion of numerous factors, which include quantity and quality-related aspects,

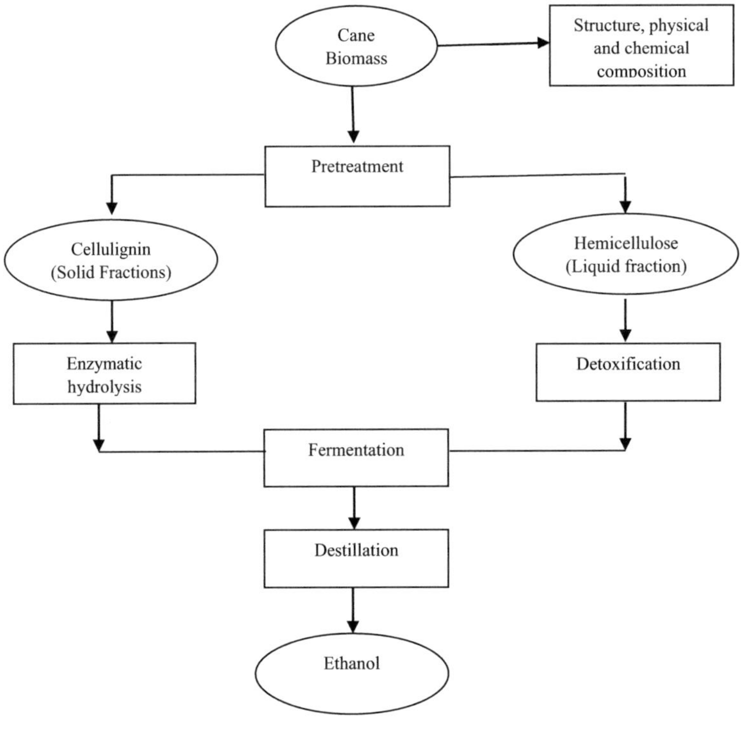

Figure 1.
Procedural flow diagram for the bioconversion of cane biomass into 2G ethanol. These renewable, high-carbohydrate raw material sources do not compete with the demand for food or animal feed. It is still difficult to market cellulosic ethanol due to the limitations in efficient bioconversion of SB/SS (efficient pre processing techniques, deconstruction of cellulosic fiber, and bioconversion of generated sugars). Extensive pre-treatment and the emergence of the effective biological conversion process, which incorporates strains of ethanol-producing bacteria that can convert pentose and hexose sugars, are among the technological hurdles that must be addressed [3–5].

estimated by various measurements, or at least the creating standard units for benchmarking, highlights the complexity of sustainability monitoring in the sugar market [7]. Sustainability success can be assessed by the improvements in productivity of sugar within the existing plant area simultaneously reducing ecological impact with socially acceptable and effective coordination between stakeholders, markets, and policies made for the public across nations which will positively impact the environmental, societal, and financial benefits [7].

Sugarcane yields one of the highest amount of sugars per hectare as a commercial crop grown in tropical and sub-tropical regions that is highly productive in harvesting sun energy, atmospheric carbon dioxide, water, and nutrients, basically Nitrogen, into biomass and simple pentoses like sucrose [2]. The production of food for human consumption, animal feed, biofuels, value added products, and highly specialized commercial products can be attained with potentially and commercially profitable raw materials. Employing various agricultural by-products and sugarcane by-products such bagasse, sugarcane tops and molasses, can lead to the production of a number of value-added products. This comprises a variety of enzymes, organic acids, amino acids, pigments, animal feed, composite, chelating agents, alkaloids, bioethanol, biodiesel, biobutanol, 2, 3-butanediol, biohydrogen, bioelectricity, and biopolymer. In addition to catering the needs of more than 80% of the sugar consumed globally in excess to 100 countries [7]. The largest producer of sugarcane in the world is Brazil (40 percent of world production). India, Thailand, China, Mexico, and Pakistan, round out the list of nations that produce sugar [2, 7].

The negative perception of refined sugar being identified as a potent causative agent in lifestyle diseases like hypertension, overweight, and insulin resistance, the lack of stability in the market of refined sugar as a commodity, the impact of increasingly popular use of alternative sweeteners like high fructose corn syrup HFCS, Stevia (*Stevia rebaudiana*), and other high intensity sweeteners (HIS) in the food industry are just a few of the obstacles that the agribusiness continues to face [2, 8].

Nevertheless, sugarcane farming, which is a subject of agro-economic research, is the most pursued research topic in the conventional linear industrial economy in cane agriculture [3]. The sugar agroindustry stakeholders are working together to fully exploit the mechanical and environmental features of biorefineries, sugar mills, and distilleries in spite of concentrating on the effects of field practices on the environment when taking the Sustainable Developmental Goals (SDGs) into account [9, 10].

2. Chemical composition and properties of sugarcane

The sugar cane plant's variety and the type of cultural management are some of the elements that impact the stem's length, which can range from less than 2 m to over 4 m in size in an adult stem; the stem's diameter changes as well, fluctuating between 250 and 350 m in the middle [4]. The color is influenced by agronomic factors, anthocyanin content, and chlorophyll content; the sugarcane material has a wide range of moisture contents, ranging from as high as 82.3 percent (tops) to as low as 13.5 percent (dry leaves); the three components of the straw have similar values for fixed carbon, volatile matterand ash with cane bagasse having a lower value for ash; virtually all substances have the same percentages of carbon (45%), nitrogen (0.5–1%), hydrogen (6%), oxygen (43%), and sulfur (0.1%) [8]. The three parts of the SS's mineral composition differ slightly for alkalis and phosphorus, showing that the SS's content grows from the tops of the dried leaves and is significantly higher than SB [4, 11].

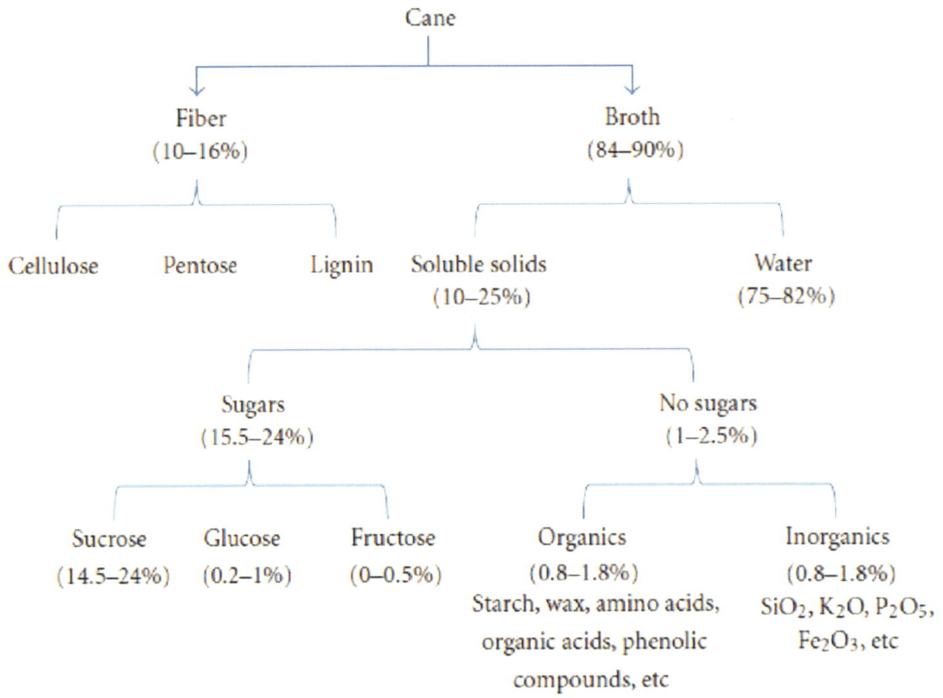

Figure 2.
General chemical composition of Sugarcane.

As per estimates, sugar cane bagasse comprises 38.9% glucan, 23.9% Lignin, 20.6% xylan, 5.6% arabinan, and 11% others. The relative compositions of major lignocellulosic fractions i.e., cellulose, hemicellulose, and lignin depends on numerous aspects, which includes plant genetics, growing environments, processing conditions, and methodologies used for the compositional analysis [12], therefore the fact that relative composition of plant polymers varies within the same type of plant material is not surprising. The composition of samples from various origins, taken by various laboratories using various methodologies, cannot be compared (**Figure 2**) [11].

3. Pre-treatments for conversion of sugarcane bagasse

3.1 Physical pre-treatments

3.1.1 Milling

A mechanical preparation called milling reduces the crystallinity of cellulose and disrupts the structure of lignocellulosic materials [13]. The most used technique is ball milling, which involves reducing the size of the particles by contacting the biomass with balls. Because this process does not require the addition of chemicals and does not produce inhibitors, it can be regarded as environmentally friendly. The high-power requirements of the equipment and ensuing high energy expenditures are a drawback of milling. For sugarcane bagasse pre-treatment, numerous cycles and miller passes are required, and the cycles often last for a longtime [13, 14].

3.1.2 Pyrolysis

High temperatures (greater than 300°C) are used during the pyrolysis process. Rapid cellulose degradation into H_2, CO, and leftover char results from this mechanism [14]. Following the removal of the char, the recovered solution is predominantly made up of glucose, which can subsequently be fermented to produce ethanol. Operating temperature, residence time, rate of heating, reaction time, type of sweeping gas, reactor type, type and amount of catalyst, and flow rate are some of the variables that affect the quality and yield of products following pyrolysis and feedstock characteristics.

3.1.3 Microwave

Pre-treatment using microwave technology is regarded as a substitute for conventional heating. Microwave pre-treatment involves application of electromagnetic field producing high heating efficiency and simple operation, in contrast to conventional heating methods that use surface heat transfer [15]. The quick times of reaction and uniform heating of the reaction mixture are this technique's key benefits.

3.2 Physicochemical pre-processing

3.2.1 Hydrothermal treatment or steam explosion

This is among the most used pre-treatment techniques for pre-treatment of substrates (also known as hydrothermal). When it comes to environmental problems, this pre-treatment involves little or no chemical input, making it a promising technology. With low lignin solubility, steam explosion treatment results in high hemicellulose solubility (generating primarily oligosaccharides). A possible method to increase the amount of fermentable sugars is the steam explosion procedure followed by enzymatic saccharification [16].

3.2.2 Ammonia fiber explosion (AFEX)

Combining steam explosion and ammonia in liquid form make up the AFEX process. It is an alkaline thermal treatment that involves rapidly releasing pressure after rapidly exposing the lignocellulosic substrate to high pressure and temperatures simultaneously. This pre-treatment can considerably increase the rate of fermentation of different grasses and herbaceous plants. The key benefits of AFEX are its effective lignin removal, low production of inhibitors, and significant carbohydrate retention in the substrates. Additionally, it is a quick and easy method [17]. The material's structure is altered during the AFEX, increasing its capacity to store water and its susceptibility to enzyme digestion of substrates (hemicellulose and cellulose), resulting in a high sugar recovery rate.

3.2.3 CO_2 explosion

In accordance to the idea that CO_2 can generate carbonic acid and speed up the pretreated material's hydrolysis, the CO_2 explosion happens similarly to the ammonia explosion. Because CO_2 explosion uses mild temperatures throughout the process and prevents any significant monosaccharide breakdown, it has relatively higher conversion efficiency than a steam explosion, is more economically feasible than ammonia explosion, and does not produce xenobiotic inhibitors. This process is safe for the environment, non-flammable, and nontoxic. It is a method with complicated process and challenging operation, nevertheless [13].

3.2.4 Treatment with hot water

With this technique, a significant portion of the hemicellulose fraction is removed while under high pressure conditions hot water hydrates the cellulose in the lignocellulosic biomass. The absence of chemicals in this process makes it unnecessary to utilize materials that resist corrosion in the hydrolysis reactor, which is one of its key advantages. Additionally, it's not necessary to reduce the raw material's size [18].

3.2.5 Acid pre-treatment

The most widely utilized acid, H_2SO_4, encourages hemicellulose hydrolysisto xylose and other sugars when it comes into contact with biomass [16, 18]. Other acids, such as HCl, nitric acid, phosphoric acid, and oxalic acid also demonstrated positive outcomes. Typically, the process can be carried out at temperatures between 120 and 180°C and residence periods between 15 and 60 minutes [19]. The acid pre-treatment method has the benefit of operating at low and medium temperatures, which reduces energy expenditures.

3.2.6 Alkali pre-treatment

Alkali pre-treatment is a delignification method that also significantly solubilizes hemicellulose. It uses a variety of bases, such as sodium hydroxide, calcium hydroxide (lime), potassium hydroxide, ammonia hydroxide, and sodium hydroxide combined with hydrogen peroxide or other substances. Compared to other pre-treatment technologies, this method uses lower temperatures and pressures, but pre-treatment takes hours or days. Alkali pre-treatment's efficacy varies with the substrate and pre-treatment circumstances in addition to severity of chemicals used. In general, lignocellulosic substrates with low lignin content such as gramineous crops, herbaceous crops and hardwoods, gives better yields with alkaline pre-treatment compared to softwoods with high lignin content [20].

3.2.7 Oxidative delignification

In this process, the peroxidase enzyme catalyzes the breakdown of the lignin in the presence of H_2O_2. A wide range of biomass, including corn stover, barley straw, sugarcane bagasse, wheat straw, rice straw, and bamboo have all undergone this pre-treatment process [21].

3.2.8 Ozonolysis

The hemicellulose and lignin fractions of lignocellulosic substrates like wheat straw, pine, bagasse, cotton straw, peanut, and sawdust of poplar can be broken down using ozone. The advantages of ozonolysis pre-treatment include the efficient removal of lignin, the lack of harmful residues for the subsequent procedures, and the fact that the reactions take place at room temperature and pressure [22]. However, the process is costly because a lot of ozone is needed.

3.2.9 Organo solvent treatment

One of the most promising approaches for pre-treating lignocellulosic materials is the Organo solvent procedure. When compared to other processes of a similar nature, the organo solvent procedure produces less waste and consumes less chemical energy to neutralize the hydrolysate. As a catalyst, substances like NaOH or Na_2SO_3 may be utilized [23]. This technique has been shown to remove lignin with great efficiency and high carbon dioxide pressure.

3.2.10 Wet oxidation

The sodium carbonate catalyst is the most common one employed in the wet oxidation process, which takes place in the presence of oxygen or catalysed air. High quantities of biomass can be converted into monosaccharides through wet oxidation, with little to no furan and phenolic aldehyde production. With an increase in aliphatic acids during the wet oxidation process, delignification is reportedly documented. This preliminary care is deemed pricey [24]. The primary benefit of this pre-treatment is its combination with alkalis, which allows released sugars to be obtained without the production of furfural and 5-hydroxymethylfurfural, two molecules that are unfavourable for fermentations.

3.2.11 Evaporation (concentration)

A physical method of detoxification that minimizes the concentration of volatile chemicals like acetic acid, vanillin and furfural, is the concentration of hydrolysate by evaporation by applying vacuum [25]. However, this procedure has the drawback of making extractives' non-volatile hazardous chemicals more prevalent.

3.2.12 Use of membranes

Membranes have a number of benefits over traditional extraction. Membrane adsorption inhibits the mixing of the organic solvent (solvent phase), which is likely to inhibit microbial growth and survival, with the aqueous phase (hydrolysate) [26]. The membranes' internal pores have functionally active surface groups linked to them, which can help eliminating the common metabolic inhibitors like acetic acid, furfural, 5-hydroxymethylfurfural (HMF), formic acid, levulinic acid, and sulfuric acid (**Figure 3**) [27].

3.3 Biological treatment

The biological technique makes use of particular enzymes or microorganisms that modify the inhibitory chemicals in the hydrolysate. Enzyme usage is a well-researched and promising technique. White rot fungi-derived laccase and peroxidase enzymes have been discovered to be efficient at removing phenolic chemicals from lignocellulosic hydrolysates. Inhibitor chemicals from lignocellulosic hydro-lysates have also been removed using microorganisms [29]. Yeasts, fungus, and bacteria are among the microorganisms that can ingest inhibitory chemicals naturally. While keeping cellulose and hemicellulose in the substrate, these bacteria can efficiently break down lignin. Another name for this process is in situ microbial delignification (ISMD). Another intriguing replacement for the detoxification process is the adaption

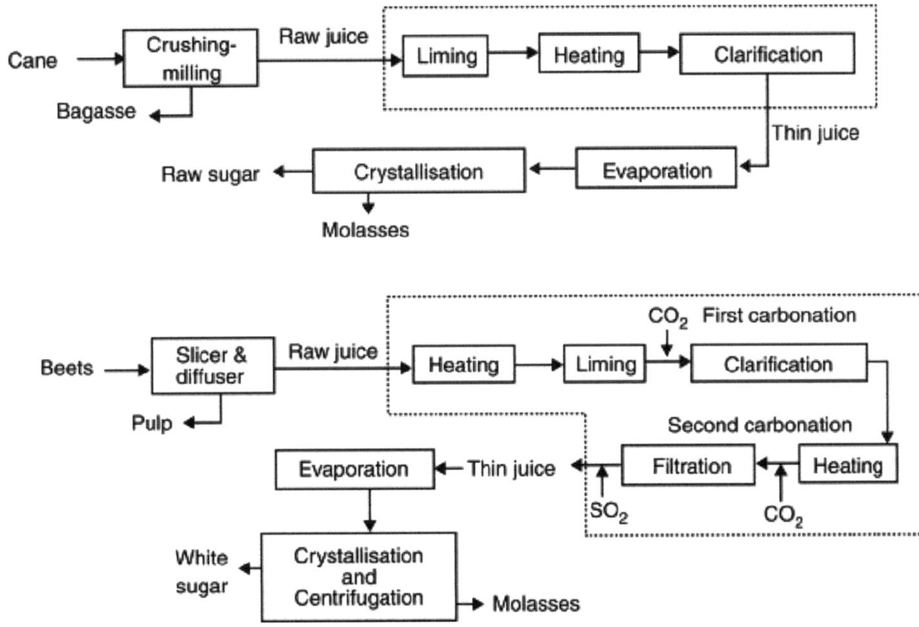

Figure 3.
Pre-treatment of sugarcane for the production of eyhanol/Biofuel [28].

of a bacterium to a non detoxified hydrolysate. This approach relies on a series of fermentation where the microbe from one experiment serves as the inoculum for the subsequent one [12]. Utilizing specialized microbes lowers the cost of detoxification and prevents the loss of fermentable sugars.

3.4 Bioconversion of hexose sugars into ethanol

Hexose sugars are bio transformed into ethanol. In the biological process of ethanol fermentation, carbohydrates are transformed by microorganisms into ethanol and CO_2. Even though there are numerous techniques and procedures for using ligno-cellulosic materials to produce ethanol, it is still challenging to produce economically viable amounts of ethanol from lignocellulosic wastes. Yeasts are the microorganism most frequently employed in the fermentation process, and Saccharomyces cerevisiae is the yeast most commonly utilized for ethanol fermentation. This yeast can thrive on the disaccharide sucrose as well as simple sugars like glucose. *S. cerevisiae* is one among the most sought-after microbes for ethanol production due to its long history in commercial fermentation processes, robust genetic transformation system, and availability (**Figure 4**) [30, 31].

3.5 Bioconversion of pentose sugars into ethanol

Pentose sugars are bio-converted into ethanol. In order to develop an affordable and practical conversion technique for the manufacture of bioethanol from sugarcane bagasse (SB) and sugarcane straw, the maximum utilization of all sugar fractions is necessary (SS). The hemicellulose fraction must be fermented at the same conversion rates as the cellulose fraction in order to get the necessary ethanol yields from Sugarcane bagasse or sugarcane straw hydrolysates. In order to absorb pentose sugars,

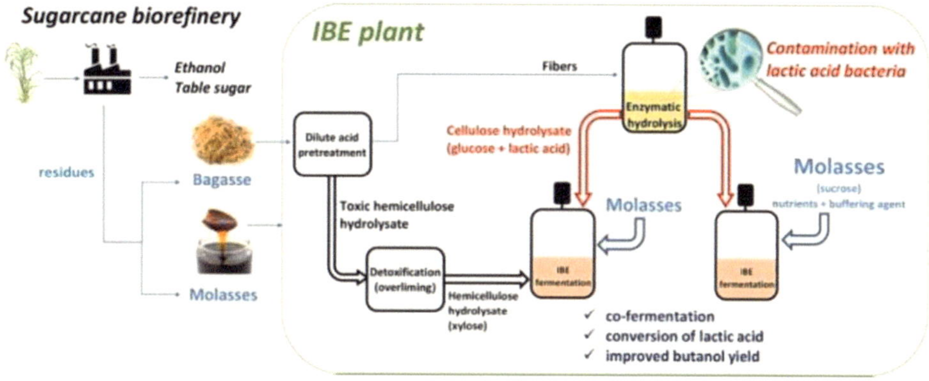

Figure 4.
Bioconversion steps for the production of ethanol from Sugarcane [32].

the enzyme D-xylose reductase (EC 1.1.1.21) converts xylose to xylitol, which is then immediately oxidized by the enzyme xylitol dehydrogenase (EC 1.1.1.9), yielding D-xylose-5-phosphate. Transaldolase, transketolase, and ribulose phosphate-3-epimerase progressively transform D-xylose-5-phosphate into fructose-6-phosphate and glyceraldehyde-3-phosphate via non-oxidative rearrangement, leading to the synthesis of ethanol by the Emden reaction -Meyerhoff Route [30, 33].

3.6 Distillation of ethanol

It is required to recover the ethanol from the fermenting broth. Ethanol (5–12 weight percent) and water make up the final medium [34]. Conventional distillation methods cannot separate the ethanol and water because they create an unfavourable mixed system. Distillation, rectification, and dehydration are the three processes in the ethanol purification process. In the first two processes, a highly concentrated ethanol solution is produced (approximately 92.4% wt), and the mixture is subsequently dehydrated to produce anhydrous ethanol. Azeotropic distillation, liquid–liquid extraction, extractive distillation, adsorption, or more sophisticated hybrid separation techniques can all be used to achieve the dehydration.

4. Sustainability in economizing the biorefinery of the sugarcane industry

The majority of sugarcane mills operating now are plants that also produce ethanol; the other plants only produce ethanol (autonomous distilleries). Through innovative framework, the so called Virtual Sugarcane Biorefinery, technical, environmental, and economic effects of these first-generation sugarcane processing facilities can be examined [35]. Optimization methods have the ability to improve economic outcomes and lessen environmental consequences for both autonomous and annexed facilities when compared to basic scenarios. Further while taking into account the average pricing over the previous 10 years, annexed facilities that diverted more sugarcane juice for the manufacturing of sugar were more lucrative. Additionally, findings show that scenarios involving more flexibility in an annexed plant are more economic. If price hikes were to occur, this alternative would be more profitable than the typical annexed one (diverting 50% of the sugarcane juice to sugar and 50% to ethanol production) [35, 36]. This helps to comprehend the true advantages of the sugarcane plant's adaptability by quantitatively demonstrating the advantages of optimization strategies.

As a consequence of the sugar beet processing, sugar beet pulp has a lot of potential as a bioeconomy component because it can be transformed into a number of useful products and by-products. The pulp is manufactured in vast quantities, it is inexpensive, underutilized, and has a desirable chemical make-up. High polysaccharides fraction is one of the beneficial components that are associated with the latter. However, it should be emphasized that the composition of sugar beet pulp might vary, depending on things like the weather. The conversion of beet pulp into bioethanol is one method for creating products with added value [37]. Comparing the use of sugar beet pulp to selling it as feed may boost its economic feasibility. Lower energy prices could have positive effects because pulp used to make biofuels does not need to be dried. Simultaneous saccharification and fermentation were employed to produce ethanol from sugar beet pulp [28].

An opportunity for both the provision of sustainable energy and the reduction of greenhouse gas emissions is provided by the biorefining of sugar beet pulp biomass for bioenergy generation. Life cycle assessment (LCA) of numerous energy-focused biorefinery scenarios and a few conversion routes were carried out to evaluate the significance of the alternative use of biomass leftovers, including sugar beet pulp (two involving bioethanol and two biogas). As an input to the LCA, specialized biochemical models are designed to establish precise mass, energy, and substance balances for each biomass conversion route [28, 32]. The results of this study generally confirmed the findings of other studies that highlighted the environmental benefits of conversion pathways including electricity and heat provision.

The standard sugar beet business was to be restructured into a revolutionary biorefinery where the sugar beet pulp can be successfully fractionated into pectins, phenolic compounds, and a sugar-rich hydrolysate, which can then be used as a fermentation feedstock for the production of succinic acid, an essential platform chemical for the growth of sustainable chemical industry and a precursor for the creation of numerous bulk chemicals, polymers, and resins [37].

Sugarcane bagasse, a waste by-product from the sugar processing industry can be used for the production of a solid fuel that has a high calorific value of around 28.2 MJ/kg using torrefaction technology [28, 32]. Due to its high fixed carbon content of 76 percent and high heating value, bio coal can be utilized in place of coal. A high calorific value bio coal with the potential to be produced from sugarcane bagasse was looked into as a potential coal substitute for the sugar industry [36]. Bagasse, a lignocellulose waste product of the sugar-processing industry, has the energy potential to make environmentally beneficial bio coal. The yield of biochar during torrefaction was 70%. The biochar can be ground to a 300-μm sand size, compressed at 2–8 MPa using molasses in a 70:30 ratio, and finally turned into bio coal. The bio coal had a calorific value of 28.2 MJ/kg, 6.3 percent moisture, 74.6 percent fixed carbon, and 1.4 percent ash. Due to the characteristics of bio coal, it can be used as an alternative to coal and reintegrated back into the sugar manufacturing sector [32, 37].

The sugarcane bio coal has a high burning period of 30 minutes and a low ignition time of 10 s. Biocoal is an eco-friendly fuel due to the low pollutants emitted during burning.

5. Conclusions

In numerous nations sugarcane, straw and bagasse (SB) are appealing second-generation renewable feedstock options. If used wisely, this feedstock might offer a steady supply of single-cell proteins, organic acids, drop-in ethanol, industrial enzymes, and other products. However, a sizeable portion of this biomass is used by businesses to generate steam and power. The remaining portion is the appropriate source of raw materials for producing high-value commodities. Due to the rapid advancements in downstream processing technologies, pre-treatment strategies, and advanced microbial biotechnology over the past three decades, it is now possible to use sugarcane wastes for the large-scale manufacturing of a variety of products without endangering the needs for food and feed. The greatest obstacle to the efficient use of these leftovers is recalcitrance/resistance of biomass to bioconversion by enzymatic fermentation. Pre-treatment is a necessary technique to improve the accessibility of carbohydrate polymers to enzymes for the subsequent hydrolysis reactions to produce fermentable sugars in order to overcome the biomass recalcitrance. There are a number of effective pre-treatment techniques available; however, the final decision regarding the pre-treatment process will depend on the efficiency of delignification or hemicellulose removal, economic viability, the minimal amount of inhibitor generation, time savings, low sugar loss, and the ability to cause the least amount of environmental pollution. Following enzymatic degradation and hemicellulose depolymerization, the sugars that were liberated are transformed into ethanol by the appropriate ethnologic strain. The ethnologic strains should be able to use pentose and hexose sugars, be resistant to inhibitors, and have good osmotic tolerance in order to produce the appropriate ethanol yields. The following six requirements are pivotal in order to establish a long-term sustainable second-generation ethanol production.

- Choosing the best pre-treatment and detoxification plan
- Development of ethanol-producing strains and in-house cellulase enzymes
- Production from pentoses like xylose, arabinose and hexose sugars that exhibit inhibitor tolerance, ethanol resistance, and quicker sugar conversion rates.
- Process optimization and intensification: combining fermentation and hydrolysis in one location.
- Quick, affordable, and efficient distillation of ethanol.
- Combining sugar/distilleries with bioethanol production facilities to share machinery, reactors, and other equipment.

Acknowledgements

Author K. Ravichandra is greatful for funding support from DBT, Govt of India and JNTU Kakinada, Andhra Pradesh.

Authors Mr. K. Rosangzuala, Ms. Rveena Gajjala, and Mr. Patlolla Ravinder Reddy acknowledge the support from AcSIR, CSIR-IICT, Hyderabad.

Dr. Linga Banoth acknowledges the support from DST-SERB EMEQ.

Conflict of interest

The authors declare no conflict of interest.

Author details

Ravichandra Korasapati[1], Suresh Babu Basthipati[2], Khawlhring Rosangzuala[3,4], Patlolla Ravinder Reddy[3,4], Raveena Gajjala[3,4], Patapanchala Ravindra Babu[1], Cheemalamarri Chandrasekhar[3], Ramesh Malothu[1], Prakasham Reddy Shetty[3,4] and Linga Banoth[3,4*]

1 Department of Biotechnology, JNTU Kakinada, Kakinada, India

2 Department of Biotechnology, JNTUH, Hyderabad, India

3 Organic Synthesis and Process Chemistry, CSIR-Indian Institute of Chemical Technology, Hyderabad, India

4 Academy of Scientific and Innovative Research (AcSIR), Ghaziabad, India

*Address all correspondence to: linga@csiriict.in

References

[1] Canilha L, Chandel AK, dos Santos S, Milessi T, Antunes FA, da Costa L, et al. Bioconversion of sugarcane biomass into ethanol: An overview about composition, pre-treatment methods, detoxification of hydrolysates, enzymatic saccharification, and ethanol fermentation. Journal of Biomedicine and Biotechnology. 2012

[2] Ptak M, Skowrońska A, Pińkowska H, Krzywonos M. Sugar beet pulp in the context of developing the concept of circular bioeconomy. Energies. 2021;**15**(1):175

[3] Manyuchi MM, Mbohwa C, Muzenda E. Evaluating the usability of bio coal from sugar canebagasse as a solid fuel. Procedia Manufacturing. 2019;**33**:516-521

[4] Formann S, Hahn A, Janke L, Stinner W, Sträuber H, Logroño W, et al. Beyond sugar and ethanol production: Value generation opportunities through sugarcane residues. Frontiers in Energy Research. 2020;**8**:579577

[5] Porter RO. Effects of the incorporation of arabinoxylans derived from selected cereals (rice bran and corn fibre) and sugarcane bagasse on the quality of baked foods: A systematic review. In: Var I, Uzunlu S, editors. A Glance at Food Processing Applications. IntechOpen. 2021. DOI: 10.5772/intechopen.99488

[6] Aguilar-Rivera N. Bioindicators for the sustainability of sugar agro-industry. Sugar Technology. 2022;**24**:651-661. DOI: 10.1007/s12355-021-01105-z

[7] Candido JP, Almeida ÉC, de Oliveira Leite DN, Brienzo M, de Franceschi de Angelis D. Vinasse from sugarcane bagasse (hemicellulose) acid hydrolysate and molasses supplemented: Biodegradability and toxicity. Ecotoxicology. 2021;**30**(5):818-827

[8] Delgadillo-Vargas O, Garcia-Ruiz R, Forero-Álvarez J. Fertilising techniques and nutrient balances in the agriculture industrialization transition: The case of sugarcane in the Caucariver valley (Colombia), 1943-2010. Agriculture, Ecosystems & Environment. 2016;**218**:150-162

[9] dos Santos JF, Canettieri EV, Souza SM, Rodrigues RC, Martínez EA. Treatment of sugarcane vinasse from cachaça production for the obtainment of Candida utilis CCT 3469biomass. Biochemical Engineering Journal. 2019;**148**:131-137

[10] Barbosa LD, Hytönen E, Vainikka P. Carbon mass balance in sugarcane biorefineries in Brazil for evaluating carbon capture and utilization opportunities. Biomass and Bioenergy. 2017;**105**:351-363

[11] de Aquino S, Fuess LT, Pires EC. Media arrangement impacts cell growth in anaerobic fixed-bed reactors treating sugarcane vinasse: Structured vs. randomic biomass immobilization. Bioresource technology. 2017;**235**:219-228

[12] Mussatto SI, Roberto IC. Alternatives for detoxification of diluted-acid lignocellulosic hydrolyzates for use in fermentative processes: A review. Bioresource Technology. 2004;**93**(1):1-10

[13] Sun Y, Cheng J. Hydrolysis of lignocellulosic materials for ethanol production: A review. Bioresource Technology. 2002;**83**(1):1-11

[14] Kumar P, Barrett DM, Delwiche MJ, Stroeve P. Methods for pre-treatment of lignocellulosic biomass for efficient hydrolysis and biofuel production. Industrial and Engineering Chemistry Research. 2009;**48**(8):3713-3729

[15] Binod P, Satyanagalakshmi K, Sindhu R, Janu KU, Sukumaran RK, Pandey A. Short duration microwave assisted pre-treatment enhances the enzymatic saccharification and fermentable sugar yield from sugarcane bagasse. Renewable Energy. 2012;**37**(1):109-116

[16] Chornet E, Overend RP. Phenomenological kinetics and reaction engineering aspects of steam/aqueous treatments. In: Focher B, Marzetti A, Crescenzi V, editors. Steam Explosion Techniques: Fundamentals and Industrial Applications. Philadelphia, PA, USA: Goran and Breach Science Publishers; 1991. pp. 21-58

[17] Sarkar N, Ghosh SK, Bannerjee S, Aikat K. Bioethanol production from agricultural wastes: An overview. Renewable Energy. 2012;**37**(1):19-27

[18] Taherzadeh MJ, Karimi K. Pre-treatment of lignocellulosic wastes to improve ethanol and biogas production: A review. International Journal of Molecular Sciences. 2008;**9**(9):1621-1651

[19] Alvira P, Tomaś-Pejó E, Ballesteros M, Negro MJ. Pre-treatment technologies for an efficient bioethanol production process based on enzymatic hydrolysis: A review. Bioresource Technology. 2010;**101**(13):4851-4861

[20] Zheng Y, Pan Z, Zhang R. Overview of biomass pre-treatment for cellulosic production. International Journal of Agricultural and Biological Engineering. 2009;**2**:51-68

[21] Yamashita Y, Shono M, Sasaki C, Nakamura Y. Alkaline peroxide pre-treatment for efficient enzymatic saccharification of bamboo. Carbohydrate Polymers. 2010;**79**(4):914-920

[22] Vidal PF, Molinier J. Ozonolysis of lignin—improvement of in vitro digestibility of poplar sawdust. Biomass. 1988;**16**(1):1-17

[23] El Hage R, Brosse N, Chrusciel L, Sanchez C, Sannigrahi P, Ragauskas A. Characterization of milled wood lignin and ethanol organosolv lignin from miscanthus. Polymer Degradation and Stability. 2009;**94**:1632-1638

[24] Carvalheiro F, Duarte LC, Gírio FM. Hemicellulose biorefineries: A review on biomass pretreatments. Journal of Scientific and Industrial Research. 2008;**67**(11):849-864

[25] Bjerre AB, Olesen AB, Fernqvist T. Pre-treatment of wheat straw using combined wet oxidation and alkaline hydrolysis resulting in convertible cellulose and hemicellulose. Biotechnology and Bioengineering. 1996;**49**:568-577

[26] Wang W, Yuan T, Wang K, Cui B, Dai Y. Combination of biological pre-treatment with liquid hot water pre-treatment to enhance enzymatic hydrolysis of Populus tomentosa. Bioresource Technology. 2012;**107**:282-286

[27] Singh R. Hybrid membrane systems-applications and case studies. Hybrid Membrane Systems for Water Purification. 2005;**3**:131-196

[28] Brienzo M, Carvalho AF, de Figueiredo FC, de Oliva NP. Sugarcane bagasse hemicellulose properties, extraction technologies, and

xylooligosaccharides production. Food Waste: Practices, Management and Challenges. 2016:155-188

[29] Anish R, Rao M. Bioethanol from lignocellulosic biomass part III hydrolysis and fermentation. In: Pandey A, editor. Handbook of Plant-Based Biofuels. Portland, Ore, USA: CRC Press; 2009. pp. 159-173

[30] Fuess LT, de Araújo Júnior MM, Garcia ML, Zaiat M. Designing full-scale biodigestion plants for the treatment of vinasse in sugarcane biorefineries: How phase separation and alkalinization impact biogas and electricity production costs? Chemical Engineering Research and Design. 2017;**119**:209-220

[31] dos Santos Vieira CF, Codogno MC, Maugeri Filho F, Maciel Filho R, Mariano AP. Sugarcane bagasse hydrolysates as feedstock to produce the isopropanol-butanol-ethanol fuel mixture: Effect of lactic acid derived from microbial contamination on Clostridium beijerinckii DSM 6423. Bioresource Technology. 2021;**319**:124140

[32] Ahmad W, Ahmad A, Ostrowski KA, Aslam F, Joyklad P, Zajdel P. Sustainable approach of using sugarcane bagasse ash in cement-based composites: A systematic review. Case Studies in Construction Materials. 2021;**15**:e00698

[33] de Assis Filho RB, Danielski L, de Carvalho FR, Stragevitch L. Recovery of carbon dioxide from sugarcane fermentation broth in the ethanol industry. Food and Bioproducts Processing. 2013;**91**(3):287-291

[34] Cavalcante WA, Gehring TA, Santaella ST, Freitas IB, Angenent LT, van Haandel AC, et al. Upgrading sugarcane biorefineries: Acetate addition allows for conversion of fermented sugarcane molasses into high-value medium chain carboxylic acids. Journal of Environmental Chemical Engineering. 2020;**8**(2):103649

[35] Troiano D, Orsat V, Dumont MJ. Status of filamentous fungi in integrated biorefineries. Renewable and Sustainable Energy Reviews. 2020;**117**:109472

[36] Buller LS, da Silva Romero CW, Lamparelli RA, Ferreira SF, Bortoleto AP, Mussatto SI, et al. A spatially explicit assessment of sugarcane vinasse as a sustainable by-product. Science of The Total Environment. 2021;**765**:142717

[37] Ribul M, Lanot A, Pisapia CT, Purnell P, McQueen-Mason SJ, Baurley S. Mechanical, chemical, biological: Moving towards closed-loop bio-based recycling in a circular economy of sustainable textiles. Journal of Cleaner Production. 2021;**326**:129325

Section 2

Sustainability and Challenges

Chapter 3

Sugar Cane Products as a Sustainable Construction Material – Case Study: Thermophysical Properties of a Corncob and Cane Bagasse Ash Panel

Rafael Alavéz-Ramírez, Fernando Chiñas-Castillo, Magdaleno Caballero-Caballero, Valentín Juventino Morales-Domínguez, Margarito Ortiz-Guzmán, Maria Eugenia Silva-Rivera, Roberto Candido Jimenez-Piñon and Angel Ramos-Alonso

Abstract

Climate change is currently an issue that worries governments and society due to its threat. It is essential to implement efficient materials with low energy consumption in construction. This work shows the use of sugarcane products in the Mexican construction sector, aiming to mitigate the impact of energy wasting. As a case study, the analysis of thermophysical properties of a light mortar panel based on cane bagasse ash and corncob is presented. The experimental thermal properties of a hybrid panel system composed of cane bagasse ash, corncob, and lime were characterized. A sandwich-type construction component was made with two outer panels of reinforced mortar and between the panel of cane and corncob bagasse ash. Measurements of the surface temperatures of the system were conducted to determine the decrement factor and thermal lag, and the results were compared to other construction systems. The decremental factor and thermal lag were 0.19 (a reduction of 82%) and 6:03 h (an increment of 2400%) compared to the control panel of ferrocement only. These results are significant because the panel prepared limits the heat flow in peak hours when high temperatures reach their maximum values. This composite panel can provide an ecological alternative for energy-saving and thermal comfort and help fight climate change.

Keywords: cane, corncob, bagasse ash, thermal properties, insulation

1. Introduction

Due to climate change, governments have become more aware of the need to save energy and provide comfort in buildings. In this context, it is essential to develop and implement bioclimatic architectural systems that reduce energy consumption without affecting the thermal comfort of the building [1]. However, it is well known that most modern buildings and houses do not adapt to changing climatic conditions. The results are energy waste, health problems, and severe environmental effects. According to Wegertseder [2], the residential sector is responsible for 40% of the planet's emissions. Forty percent of greenhouse gas (GHG) emissions and one-third of black carbon emissions are from construction industries [3, 4].

On the other hand, green building analysts predicted that, by 2030, the building and construction sectors would produce more than 40 billion tons of carbon emissions [5, 6]. Imported commercial materials such as polyurethane-based insulating foams and polystyrene used to insulate the building negatively affect the ecological environment from production to disposal as waste material. The bioclimatic design provides criteria in buildings to help reduce energy demands and pollutant emissions into the atmosphere. The thermophysical characteristics of building materials are essential in minimizing thermal gains by conduction and radiation and obtaining thermal comfort conditions and energy savings. That is one of the most immediate ways to significantly reduce emissions [7] and up to 30% of energy consumption [8].

Thermal inertia is a necessary thermophysical property for extreme climates with significant thermal oscillations during the day and night. Thermal inertia represents the ability of a material to conduct and store heat, which is related to thermal conductivity and volumetric heat capacity. In contrast, heat transfer in the building is related to how fast or slow the interior temperatures reach the exterior temperature. Researchers are interested in developing composite materials with natural fibers, which generate thermal lag when combined with thermal inertia [9–15].

Sugar cane is a grass plant, and its stalk is fibrous. It is cultivated in several countries; It becomes an agricultural residue once the juice is extracted. The sugarcane bagasse has been used in thermal and acoustic insulation. The different procedures used to extract fibers can influence the chemical composition and fiber structure [16]. The sugarcane bagasse fiber surface can be changed to increase interfacial interactions considering the environmental criteria and produce functional components from biodegradable wastes [17]. The thermal conductivity of sugarcane bagasse fiber as an insulating material has also been investigated [18]. The results showed a thermal conductivity of 0.04610 w/m·K in the average temperature range of 15.6–32°C. Aminudin [19] investigated the improvement of cement bricks by 10% weight of sugarcane bagasse fiber, decreasing thermal conductivity to 0.62 w/m·K. Some studies report that the chemical composition is affected by fiber age, harvesting method, and climatic conditions such as cellulose, hemicellulose, lignin, and sugar ash [20].

Standard ASTM 618–00 defines pozzolans as "siliceous or alumino-siliceous materials which have little or no cementitious value, but when finely divided, and in the presence of water, react chemically with calcium hydroxide $Ca(OH)_2$ at room temperature to form compounds with cementitious properties". Chemical, physical and mechanical methods can be applied to characterize the pozzolanic reactivity of a material. The mechanical methods assess the role of the pozzolanic reaction in developing mechanical compressive strength in pozzolan-containing mortars and cement concretes. The pozzolanic reactivity indicates the pozzolan-lime reaction determined by several methods [21]. Studies such as the one conducted by [22] have shown that these agricultural waste by-products contain high pozzolanic reactivity. The main components that govern the reactivity of a pozzolan are (SiO_2, Al_2O_3, Fe_2O_3, and CaO). Payá [23] evaluated the efficiency of two siliceous pozzolans, silica fume (SF) and rice husk ash (RHA), and a metakaolin silico-aluminous pozzolan MK. The results showed the high reactivity of these materials acting as pozzolans when combined with lime and the apparent dependence on the water/cement ratio of the mix and curing time.

When sugarcane bagasse is burned, sugarcane bagasse ash (SCBA), a mineral residue rich in silica and alumina, is produced. Its structure depends on the combustion temperature [24]. This product can be used as a pozzolan in cement-based pastes. Several authors [25–28] have found that using SCBA as a partial replacement for portland cement improves the durability and mechanical properties of cementitious materials. The benefits developed by SCBA are due to physical and chemical effects linked to its ability to provide amorphous silica that will react with $Ca(OH)_2$ in water during cement hydration. SCBA is usually obtained under uncontrolled burning conditions [27]. Thus, the ash may contain black particles due to the presence of carbon and crystalline silica when burning occurs at high temperatures (above 800°C) or for prolonged times. The ash quality can be improved by controlling temperature, heating rate, soaking time, and atmosphere, as previously reported for the highly pozzolanic rice husk ash (RHA) [29]. On the other hand, carbon and unburned material can lower the pozzolanic activity of sugarcane bagasse ash [25]. Particle size, calcination temperature, amorphous structure, and chemical composition all affect the pozzolanic activity of bagasse ash [30–32]. The calcination temperature at 600°C is essential for producing sugarcane bagasse ash with pozzolanic activity [31]. It presents amorphous silica, low carbon content, and high specific surface area [23, 27]. Laboratory tests on the pozzolanic reactivity of Cuban sugar cane wastes (straw and bagasse) revealed that the ashes have intense pozzolanic activity and can be used as active additives in cement manufacture [32].

1.1 Recent reserch on sugarcane bagasse ash

This section highlights recent research using sugarcane bagasse ash as a pozzolanic material. Nengsen Wu [33] developed ultra-high performance green concrete (UHPC) with sugarcane bagasse ash (SCBA) as a replacement for cement. SCBA's effects on UHPC flowability, setting time, compressive strength, and shrinkage were investigated. The results showed that using SCBA in UHPC as a cement replacement maintains compressive strength, improves workability, and reduces shrinkage of the UHPC paste. Autogenous shrinkage was reduced by 24.48%, but compressive strength was nearly identical. Yadav [34] conducted an experimental investigation to demonstrate the effectiveness, availability, and cheap cost of geopolymers generated from mechanical milled sugarcane bagasse ash and metakaolin. According to this investigation, the sugarcane bagasse ash and metakaolin need mechanical and chemical treatment to increase their pozzolanic reactivity. Mehrzad et al. [35] fabricated and tested fibrous sugarcane bagasse (SBW) samples with different densities, thicknesses, surface morphology, and tensile properties. They conclude that sugarcane bagasse fiber samples can be a new sustainable building material in terms of thermal and acoustic qualities. Brito [36] studied sugarcane and bamboo-based particleboards. Three proportions of blends (25, 50, and 75%) were adopted for particle boards. The boards made with 75% bamboo particles and 25% sugarcane bagasse particles achieved the values required by the standard for thickness swelling in 24 h.

Gharieb [37] studied the use of carbonation and lime wastes from sugar beets (CLR) to partially replace cement as a cementitious material. The optimal level of CLR was 5%, which increased compressive strength and microstructure. The results confirmed that it is possible to use CLR as a cementitious material. Jagadesh [38] explored the use of Processed Sugar Cane Bagasse Ash (PSCBA) in various proportions in cementitious mortar. Portland cement with PSCBA lowers energy usage, reduces domestic gas emissions, and enhances cement characteristics. It was observed that cement mortar's mechanical and fracture properties with 10% substitution of PSCBA in ordinary Portland cement show improved properties. Jittin & Bahurudeen [39] examined the rheological performance and compressive strength of sugarcane bagasse and rice husk ash using a ternary-based hybrid cementitious system. The yield strength, plastic viscosity, and consistency index increase with added sugarcane bagasse ash and rice husk ash. The highest compressive strength was observed for ternary mixed concrete with 10% bagasse ash and 5% rice husk ash, followed by binary concrete mixed with 20% bagasse ash. SCBA as a supplementary cementitious material (SCM) and supplemental filler material (SFM) for usage in the construction sector has awakened the interest of researchers. Processed SCBA has improved characteristics compared to its unprocessed counterpart, with an optimum replacement of 20% [40]. Therefore, it is very effective as a supplementary binder and filler material.

Souza [41] investigated the development of lightweight mortars from sintering SBA. Two different SBA were evaluated; one was produced in a ball mill and the other in a knife mill. The findings show that lightweight aggregates based on SBA and RC binary mixtures can be produced in a wide range of particle densities from 1.03 to 1.67 g/cm^3. Subedi [42] investigated the feasibility of using high contents of co-processed sugarcane bagasse ash (PBA) as a partial cement replacement for the development of engineered cementitious composites (ECC) with a 1.5% volume of polyvinyl alcohol (PVA) fibers. The results showed that the workability of the mixtures decreased with increasing PBA content. Furthermore, adding PBA decreased compressive strength up to 39% and increased surface resistivity, where composites outperformed control composites. The use of PBA produced a reduction in the crack tip.

Klathae [43] investigated bagasse ash (BA) from processed sugar mills. HVGBA concrete compressive strength, tensile strength, elastic modulus, and drying shrinkage were investigated [43, 44]. The results showed that using HVGBA could reduce the maximum temperature rise between 8 and 19°C of the control concrete (CT). All concrete incorporating HVGBA with different LOI had higher drying and shrinkage than CT concrete, increasing the LOI and cement replacement rate. Barbosa [45] examined the effect of SCBA with various chemical and mineralogical compositions on paste hydration, compressive strength, and autogenous shrinkage of mortars. SCBA samples with high levels of amorphous silica-rich, fewer pollutants, and a high specific surface area behaved like RHA, with effects on hydration. Combined SCBA mortars, regardless of SCBA type, caused an increase in compressive strength. However, only the SCBA plus pozzolanic mortar was comparable to RHA.

Athira and Bahurudeen [46] compared the microstructure of processed and received rice straw ash to SCBA. Microstructural analysis revealed that rice straw ash and SCBA are made of phytoliths, dumbbell-shaped silica storage structures. Rice straw ash is richer in phytoliths than SCBA; however, the prismatic structures in

SCBA are absent in rice straw ash. It is found that the yield strength and viscosity of the cement paste increase with the addition of rice straw.

Klathae [44] formally demonstrated that SCBA with proper particle size is an excellent pozzolanic material for concrete durability. They investigated high-strength concrete (HSC) with a high volume of sugarcane bagasse ash (HVSCBA). The results showed that the 28-day compressive strength of the binary and ternary binders HSC could be developed to meet the requirement of 55 MPa.

The following case study shows the thermal properties of a biodegradable hybrid panel based on a lightweight mortar combining sugarcane bagasse ash, corn stover, and lime. SCBA was combined with lime and generated a binder material according to the standard ASTM 618–00. A sandwich construction component was made with panels, two outer layers of reinforced mortar, and the SCBA and corncob panel (2F + POCE) in the middle. Measurements of the surface temperatures of the system were conducted to determine the thermal damping and thermal lag. These results were compared to other construction systems.

2. Case study: THERMOPHYSICAL properties of a sugarcane bagasse ash and corncob-based panel

2.1 The climate of Oaxaca city and bioclimatic strategies

The city of Oaxaca is located in the southeast of the Mexican Republic; its bioclimate is temperate [47]. Summers are humid along the eastern lowlands and present an average daytime temperature ranging from medium-low to medium-high 9 to 34°C. It has an altitude of 1550 meters above sea level and geographical coordinates of 17°04′04”N latitude, 96°43′12”W longitude. This place exhibits moderate relative humidity and summer rains as classified by Köppen [48]. Winds constantly move north to south, and solar radiation is intense on clear days. **Figure 1** summarizes the climatic data for the locality according to average values for the last decade from the National Meteorological Service (SMN). However, when the range of annual oscillation is more than 14°C, it is considered very extreme, according to Köppen-García [49].

2.2 Materials used in the hybrid panel

The SCBA used to elaborate the biodegradable hybrid panel was obtained directly from a sugarcane mill in Ciénega, Zimatlán de Álvarez Oaxaca, 20 minutes away from the capital city of Oaxaca-Mexico. **Table 1** shows a typical chemical composition of sugarcane bagasse [52]. The ash selected for this study is derived from making sugar cane, where the sugar cane juice must be squeezed, and the bagasse obtained is used to burn in the oven. The temperatures reached inside the oven are above 700°C.

The ashes obtained from the furnace’s interior were homogenized by quartering based on the Mexican Official Standard NOM-AA-61. The SCBA particles were washed and exposed to the sun for 48 hours. Selected SCBA and corn stover material were taken to the drying area at CIIDIR IPN Oaxaca. The corncob was obtained from the local market. The material used as a binder in the panel matrix was pine resin obtained from the town of Ixtlán Oaxaca, located 30 minutes from the capital city of Oaxaca-Mexico.

	Annual	March
Mean maximum	30.4	32.9
Mean minimum	13.6	12.9
Mean	22	22.9
Minimum temperature	7.0	9.7
Thermal oscillation	16.8	20
Mean wind speed	1.8	1.9
Horizontal global radiation	682.5	783
Relative humidity	65.6	55
Heating degree days (18°C)	3033.8	288.3
Cooling degree days (20°C)	- 1270.2	- 176.7

The geographic location of Oaxaca city

Figure 1.
Average climatic data and location for the city of Oaxaca, Mexico.

Cellulose (%)	Hemicellulose (%)	Lignin (%)	Ash (%)	Extractives	References
51	30.10	12.50	2	—	[50]
43.8	28.60	23.50	1.30	—	[51]

Table 1.
Chemical composition of sugar cane bagasse.

Granulometry tests were carried out on the SCBA. The sugar cane husk determines the percentage of fines in the SCBA. It ensures its pozzolanic reaction with the calcium hydroxide to determine the particle size of the husk to be used in the hybrid panel. The particle size test was carried out by adopting the recommendations of the manual M.MMP.1.103, disaggregated drying, and quartering of samples. The granulometry of both materials was carried out in a WS TYLER Sieve ROTAP machine Model RX-29 for 5 minutes. The passage of the SCBA material was done through different meshes No. 4 (4.76 mm), No. 10 (2.00 mm), No. 20 (0.84 mm), No. 40 (0.42 mm), No. 60 (0.25 mm), No. 100 (0.149 mm) and No. 200 (0.074 mm) to obtain the retained mass in each mesh, and calculate its percentage. The material that passed mesh # 200 was the percentage of fines.

In order to determine the granulometry of the corncob, it was decided to use the corncob particles retained in mesh No. 4 (4.76 mm). It was mixed with the SCBA as a cementitious agent.

	CAL	SCBA
Element/compound	%	%
Al_2O_3	N.D	9.92
CaO	68.63	2.59
Fe	0.4	2.7
Fe_2O_3	0.14	2.32
FeO	0.39	1.39
K_2O	0.22	2.1
MgO	0.42	1.44
MnO	ND	0.14
Na_2O	ND	0.23
P_2O_5	ND	0.9
PXC at 950°C	29.84	24.15
SiO_2	0.31	51.66
TiO_2	ND	0.74
Density (kg/m^3)	21.97	21.48
ND = Undetected		

Table 2.
Chemical properties of lime and SCBA.

Table 2 summarizes the chemical properties of the SCBA and lime. It was observed that 51.66% of silicon oxide was present in the SCBA, which allowed it to react with the hydroxide in lime with 68.63%.

Finally, the reinforced mortar (ferrocement) is the construction element used to cover the biodegradable panel as a sandwich type. Ferrocement is a construction material composed of reinforced concrete and several layers of reinforcing mesh, electro-welded mesh, and chicken wire, uniformly distributed throughout a cross-section (ACI 549R-93). A mortar rich in cement, sand, and water was used for its making. This material has a thickness of 0.025 m. It is characterized by its high strength and flexibility and is a low-cost material. The mortar used to make the ferrocement slab is a Portland cement type 1 from Cooperativa La Cruz Azul SCL that meets all the ASTM C-150-89 standard requirements. Natural sand, clean and free of organic substances, sieved with the # 8 (2.38) ASTM. The average grain size was 0.7 ± 0.145 mm. Water from the distribution network was taken to the locality to prepare the mixture. The mechanical properties of the mortar used in this study can be found in [53].

2.3 Preparation of test panels

A lightweight mortar made of SCBA, corncob, and lime was used as a base to elaborate the (2F + POCE) panels, whose mechanical properties were previously studied by the authors of the current chapter [54]. A lightweight mortar was also developed, taking advantage of the SCBA and lime characteristics to obtain a cementitious agent to stabilize the mixture. The SCBA and lime mixture was prepared manually; water was added until a uniform consistency was achieved. The specimens were compacted manually with a wooden mallet. Finally, two panels of dimensions 1 m x 1 m long x 10 cm thick were made to determine the thermal damping properties and thermal conductivity. The specimens were left to dry for three weeks before their thermal properties were obtained. Finally, two 1 m x 1 m x 0.25 m thick ferrocement plates were fabricated to form the hybrid sandwich panel. The two ferrocement plates cover the plate filled with a 1 x 1 x 0.10 m thick waffle matrix, forming the composite panel. **Figure 2a** shows the panel with the sugarcane bagasse-lime-corncob ash matrix. **Figure 2b** shows the hybrid sandwich panel with two layers of reinforced mortar (2F + POCE).

2.4 Determination of decrement factor and thermal lag

In order to determine the damping and thermal retardation properties, first, the analysis date had to be defined. In doing so, the data of the highest solar radiation of the locality was considered, which corresponded to March (**Figure 3**); such data were obtained from the climatic normals for Oaxaca city (1980–2010) (SMN).

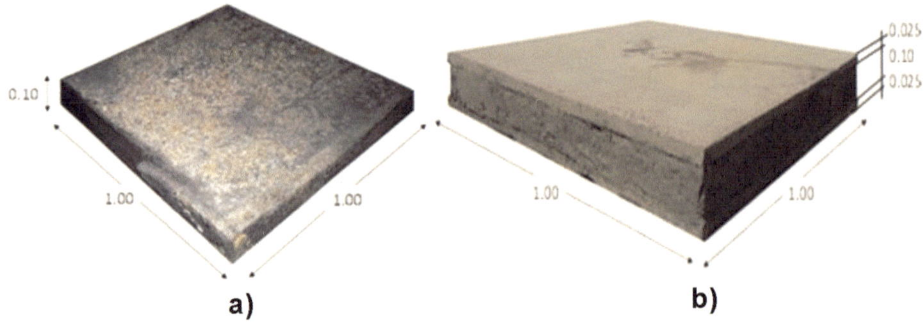

Figure 2.
Hybrid panel: a) cane bagasse ash-lime-corncob matrix, b) panel with two ferrocement plates and, in between, a panel with sugarcane bagasse-lime-corncob ash matrix.

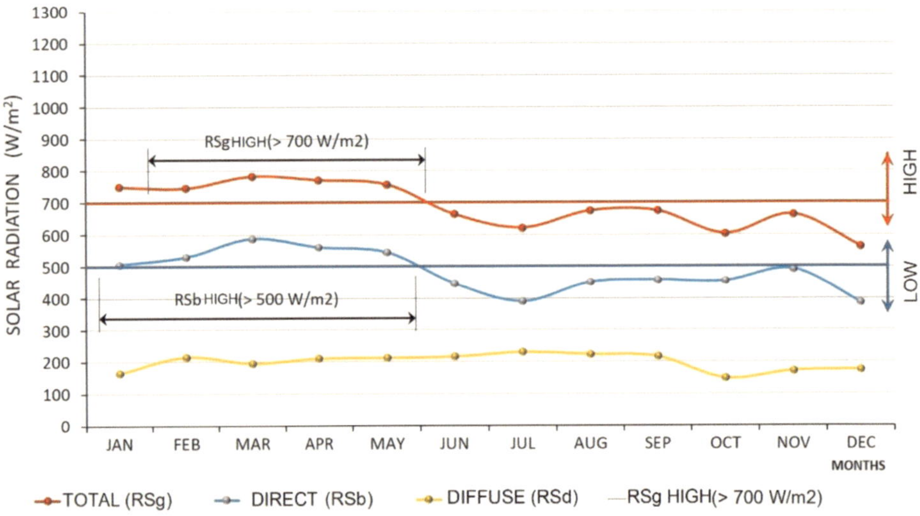

Figure 3.
Solar radiation for the city of Oaxaca.

Temperature measurements were made using five experimental chambers built for this purpose. Surface temperature measurements were carried out using 30 TMC6-HD Smart sensors (−40 to 100°C) connected to a 12-bit HOBO12 data logger system. It includes HOBOware software and a standard NIST calibration kit and stores 43,000 readings at a sampling rate of 1 s-18 h. **Figure 4** shows the placement of sensors in a thermal chamber; three were connected on the inner surface part of the specimens and three on the outer surface part of the specimens. The hobo was placed at 2.50 m inside a thermal shelter for outdoor air temperature recording to avoid direct solar radiation and allow cross ventilation.

The thermal properties of the proposed panel (2F + POCE) were compared to other construction elements built for that purpose, as shown in **Table 3**.

Figure 5a shows the placement of the different construction elements in the thermal chambers. It is worth mentioning that the thermal chambers were insulated on five sides. The upper part was left open to place the panel and allow direct solar radiation (**Figure 5b**).

2.5 Determination of thermal conductivity properties

Therm conductivity measurements were made on a homemade hot plate conductivimeter at CIIDIR IPN Oaxaca on the PO + CE panels utilized in this work [55]. This device is essentially a box with one side open, producing heat using an electric heating system. A Fiberfrax Blanket ceramic fiber quilt material resistance of 1260° is used to isolate the box's walls thermally. The conductivimeter is a device that

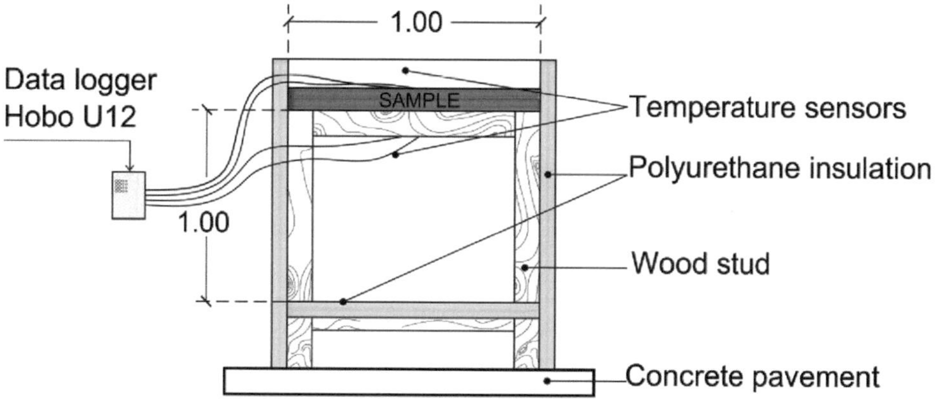

Figure 4.
Position of thermocouples, cross-section.

Wall panels	Dimensions (m)
Corncob Panel (PO + CE) (Matrix of SCBA, lime and corncob)	1,1,0.10
Ferrocement panel (LF)	1,1,0.025
Ferrocemento panels (0.025 m) + corncob panel as infill (2F + POCE) (0.10 m).	1,1,0.15
Concrete slab (CLC) (0.10 m)	1,1,0.10
Two ferrocement slabs + air (2F + A)	1,1,0.10

Table 3.
Dimensions of panels.

Figure 5.
a) Cane bagasse ash panel placement; b) general view of the thermal chambers.

employs the steady-state conduction heat transfer concept and enables standard ASTM C 177 thermal conductivity determination using the equation:

$$k = \frac{Q}{A(\Delta T/L)} \tag{1}$$

where Q represents the rate of heat flow through the specimen in W, k represents its thermal conductivity in W/m K, ΔT represents the temperature differential through it in K, L represents its thickness in m, and A represents its cross-sectional area in m^2.

It is important to note that if the sample is a compound with pores or spaces where heat can be transmitted via conduction, radiation, or convection, k in Eq. (1) represents the apparent thermal conductivity.

The upper part of the conductivity meter is uncovered, where the panel was placed. Inside, there are installed 500 watts electrical resistance of stainless steel at 127 volts single-phase with dimensions of 1 m × 1 m with flexible terminals of 0.5 m in length, and high-temperature cable with a thickness of 4 mm.

The (2F + POCE) building component temperatures were recorded utilizing TC6-J type J thermocouples with a 2 m connector that supports temperatures of 0–800°C. These sensors are connected to a four-channel HOBO UX120-014 M data acquisition system, which stores the temperature information recorded by the thermocouples. **Figure 6** shows the flow diagram of the conductivity meter.

For this investigation, the PO + CE panel wall underwent five testing replicates. Temperature readings were obtained every 10 minutes. Only the range of steady state was taken to determine the mean temperature of each sensor after registered temperatures from the four sensors on the hot side and the four sensors on the cold side. The four sensors on the hot side were then averaged after that. The four sensors on the cold side underwent the same process. In order to determine the thermal conductivity of the panel wall tested, the temperature gradient was computed as the hot-side average temperature less than the cold-side average temperature. It should be noted that the mean temperature recorded in each sensor for each panel wall was calculated from more than 100 steady state points.

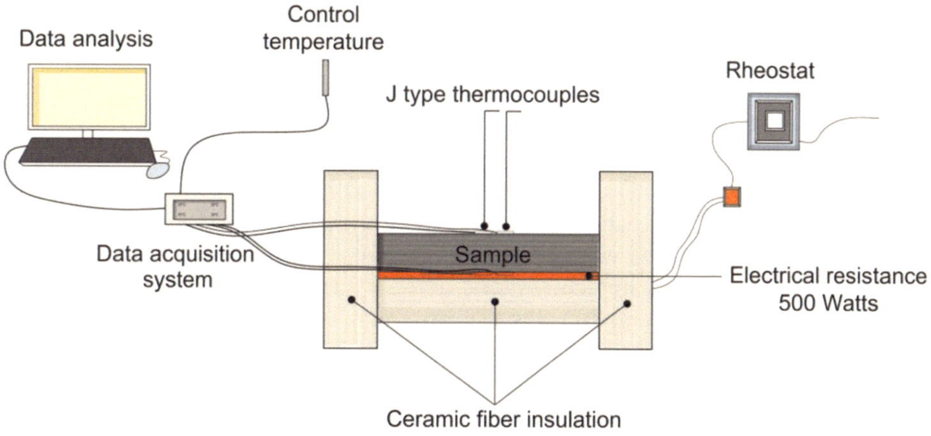

Figure 6.
Thermal conductivity test flow chart.

3. Analysis and discussion of results

3.1 Bioclimatic strategies for Oaxaca City

The results of the psychrometric chart (**Figure 7**) show that 77% of the time, people are thermally uncomfortable, and only 22.4% have thermal comfort conditions in the city of Oaxaca. It is observed that there are requirements of internal heating gain of 33.5%, thermal mass of 26.7%, shading of 22%, adaptive comfort ventilation of 18.6%, and forced ventilation of 21.9% for cooling. It is essential to mention that the percentage of internal heating gains (33.5%) is generated with the thermal mass

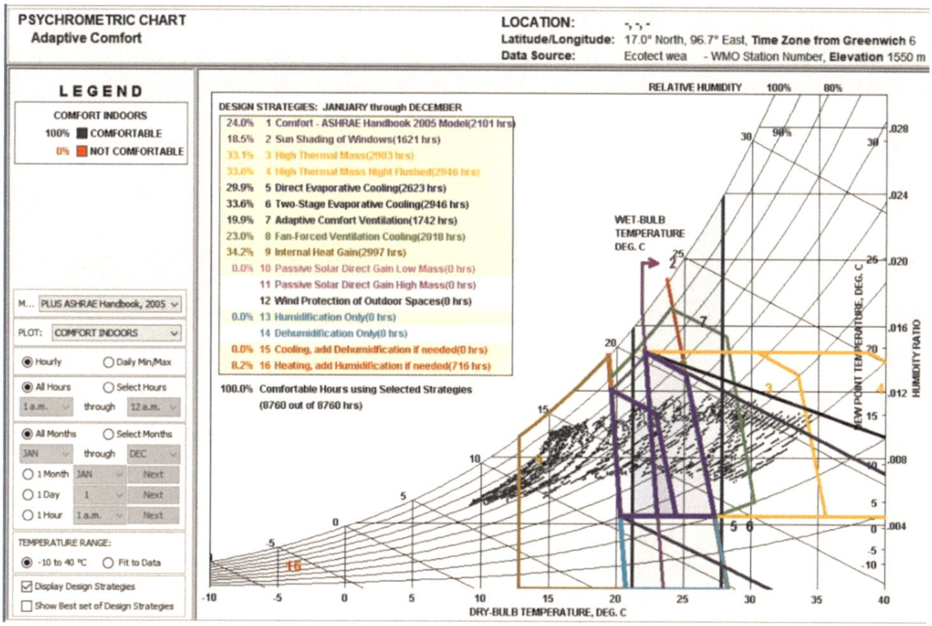

Figure 7.
Psychrometric chart for the city of Oaxaca.

because it can store the energy required during the day for passive heating. In this sense, the proposed hybrid panel can be a viable ecological alternative.

3.2 Results of thermal lag and decrement factor

Figure 8 presents the thermal measurements monitored inside the experimental cells taken in March. The analysis period was determined from March 21 to 24, 2014, because they presented the most stable temperatures.

The temperatures evaluated correspond to the values of the maximum temperatures (Tmax), minimum temperatures (Tmin), mean temperature (Timed), and the thermal amplitudes of temperature presented on the outer and inner surface of each panel. **Figure 9** shows the evolution of the exterior surface temperature wave, while **Figure 10** shows the interior surface temperatures recorded. It is observed that the highest values of thermal amplitudes were presented in the exterior surface temperatures due to the direct exposure to solar radiation. It is observed that the highest exterior thermal amplitude was presented in the (PO + CE) specimen because it presented a very high surface temperature attributed to the panel's black color. Regarding the interior surface temperatures these were presented as follows:

$$(2F + POCE) < (PO + CE) < (2F + A) < (CLC) < (LF) \quad (2)$$

Table 4 summarizes the results of the thermal amplitudes obtained on the exterior and interior surfaces of the specimens. The decrement factor of the analyzed roof specimens is also presented. **Figure 10** shows the results for the interior thermal amplitudes, which were given as follows (2F + POCE) < (PO + CE) < (2F + A) < (CLC) < (LF). **Figure 11** shows the decrement factor from smallest to largest following the same pattern: (2F + POCE (0.19)) < (PO + CE (0.30)) < (2F + A (0.50)) < (CLC (0.73)) < (LF (1.06)), this indicates that the specimen (2F + POCE) achieves the lowest thermal decrement factor of external temperature waves.

Table 5 shows the comparative results of the average thermal lag obtained from the indoor surface temperatures concerning the outdoor surface temperatures of the roof systems under study. The average thermal lag values obtained from the 4 days of the experimental period (March 21 to 24) are presented from highest to lowest and were given as follows: (2F + POCE (6:03)) > (PO + CE (3:52)) > (2F + A

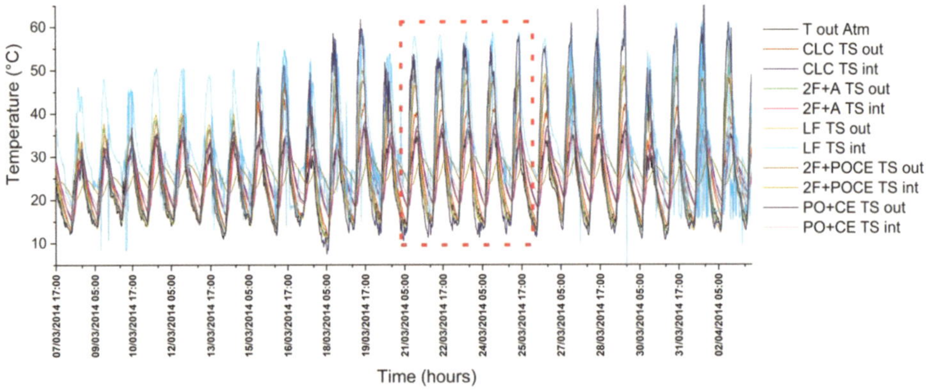

Figure 8.
Indoor ambient + outdoor ambient temperatures (March 7 to April 3).

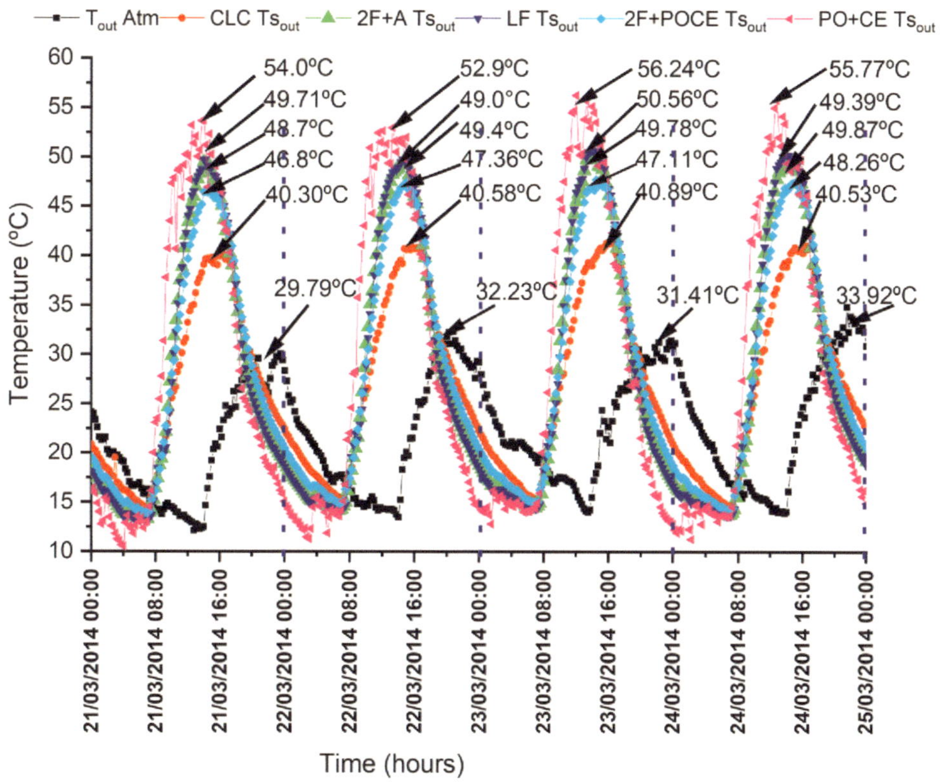

Figure 9.
Thermal amplitudes based on outdoor surface temperature in (2F + POCE) vs. (PO + CE) vs. (2F + a) vs. (CLC) vs. (LF) roofs and outdoor ambient temperature.

(2:30)) > (CLC (0:30)) > (LF (0:15)). **Figure 12** exemplifies the thermal lag obtained for March 23. The results obtained from the average thermal lag indicate that the specimen (2F + POCE) achieves the highest thermal lag of the outside temperature waves (6,00 hours).

Belhadj et al. [56] investigated straw and cement walls with different thicknesses from 0.1 to 0.25 m reporting thermal lag values of 4 to 8 hr. and a decrement factor of 0.15 to 0.85. Panel 2F + POCE presented a thermal lag of 6:00 hrs and a decrement factor of 0.19. These values are thermally better than the compressed earth specimens by Roux-Gutiérrez and Velázquez-Lozano [57] with different types of plasters. The double block compressed earth without plaster recorded a thermal lag of 4:15 hrs and a decrement factor of 2.28. Other conventional materials such as baked mud brick or cement mortar block recorded thermal lag values of about 0:30 hr. and decrement factor values of about 1.88. Gallegos-Ortega et al. [58] studied a house with thatched walls of 0.4 m in width in Tecate, Baja California, Mexico. Their results showed a thermal lag of 9 to 12 hrs, which means 2 hrs above the 2F + POCE panel, influenced by the greater thickness of the specimen.

The control panel LF, made of ferrocement only, shows the lowest thermal lag and highest decrement factor compared to the other panels tested. CLC panel showed a thermal lag increment of 200% and decrement factor reduction of 31%; panel 2F + A had a thermal lag rise of 1000% and a decrement factor reduction of 52.8%; panel PO + CE had a thermal lag rise of 1360% and a decrement factor reduction of 71.7%. The proposed panel 2F + POCE showed a thermal lag increment of 2400% and a decrement factor reduction of 82%.

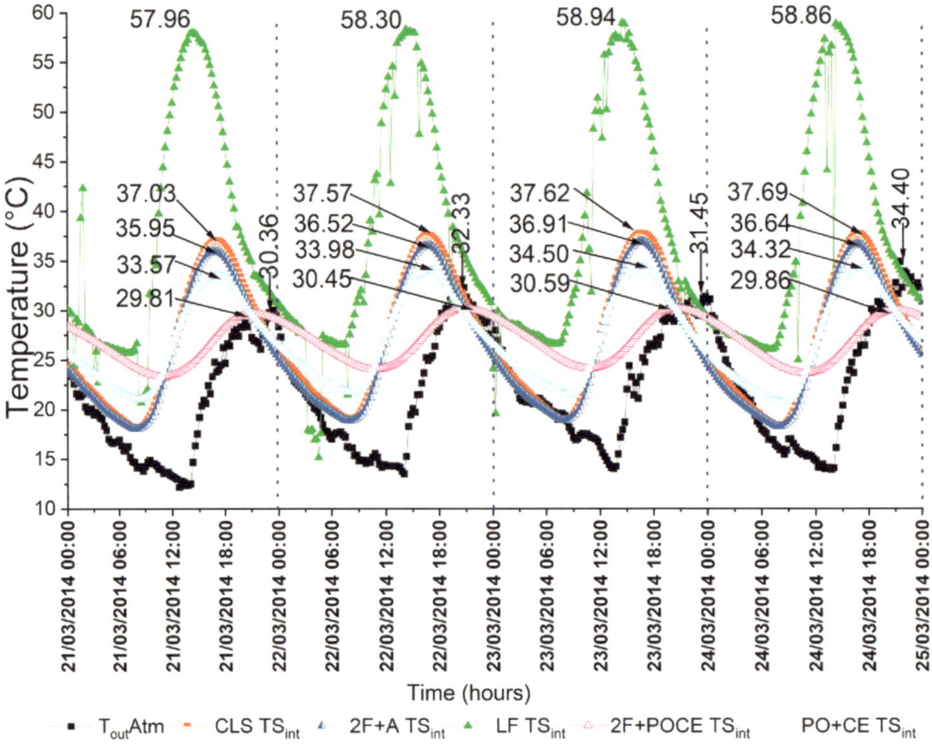

Figure 10.
Thermal amplitudes based on indoor surface temperature in (2F + POCE) vs. (PO + CE) vs. (2F + a) vs. (CLC) vs. (LF) roofs and outdoor ambient temperature.

Specimen	Outside surface temperature from March 21 to 24			Inside surface temperature from March 21 to 24			Decrement factor
	Temp Max (°C)	Temp Min (°C)	ΔT Outside surface	Temp Max (°C)	Temp Min (°C)	ΔT Inside surface	
LF	50.20	13.81	36.39	58.52	20.10	38.42	1.06
CLC	40.92	14.94	25.98	37.52	18.65	18.87	0.73
2F + A	49.68	13.89	35.79	36.53	18.52	18.01	0.50
PO + CE	54.66	11.47	43.20	34.12	21.27	12.86	0.30
2F + POCE	47.26	14.49	32.77	29.99	23.76	6.24	0.19

Table 4.
Thermal amplitude values and decrement factor of (2F + POCE) vs. (LF) vs. (CLC) vs. (2F + a) vs. (PO + CE).

Thus, the results obtained in this work with the proposed panel 2F + POCE indicate that it is a viable option for use in walls and roofs in warm and temperate climates with high thermal oscillations.

3.3 Results of thermal conductivity

In the conductivimeter, a heat flow of 32.4 watts was fed to the panel PO + CE. The thermal temperature in a steady state was monitored for 4 days. The mean thermal

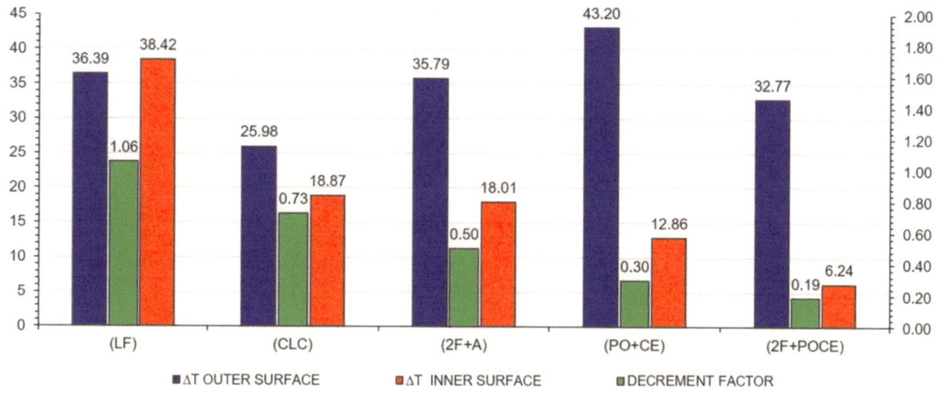

Figure 11.
Comparative results of the decrement factor in roofs: (2F + POCE) vs. (PO + CE) vs. (2F + a) vs. (CLC) vs. (LF).

Thermal lag mean from march 21–24, 2014					
Experimental cells	**21**	**22**	**23**	**24**	**Thermal Lag**
Sample 1 (2F + POCE)	6:15	6:00	6:00	6:00	6:03
Sample 2 (PO + CE)	3:00	3:45	4:45	4:00	3.52
Sample 3 (2F + A)	2:45	1:45	2:30	3:00	2:30
Sample 4 (CLC)	0:30	0:45	0:15	0:30	0:30
Sample 5 (LF)	0:00	0:15	0:30	0:30	0:15

Table 5.
Comparative results of thermal lag in roofs.

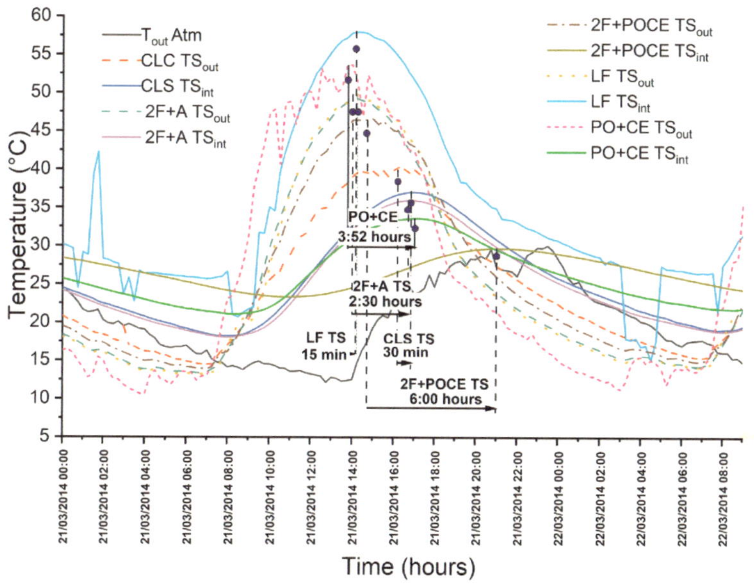

Figure 12.
Comparative results of thermal lag in roofs: (2F + POCE) vs. (PO + CE) vs. (2F + a) vs. (CLC) vs. (LF).

conductivity of panel PO + CE was 0.2008 ± 0.0064 W/m·K, Maximum 0.207 W/m·K, minimum 0.194 W/m·K, median 0.203 W/m·K.

Insulating materials have thermal conductivity values of 0.01 to 0.09 W/m K. Alavez-Ramirez et al. [55] found thermal conductivity values for ferrocement of 0.69 W/m K, ferrocement plus coconut fiber 0.221 W/m K, lightweight concrete brick 0.53 W/m K, hollow concrete block 0.68 W/m K and red clay brick 0.93 W/m K. Ruiz Torres et al. [59] reported thermal conductivity for thermos-slab 0.263 W/m K, cement-sand mortar 0.63 W/m K, typical annealed red brick 0.872 W/m K. Borbón Almada et al. [60] reported a thermal conductivity for annealed mud brick 0.814 W/m K, lightened block 0.465 W/m K, cement-sand mortar 0.470 W/m K and recycled cement-sand mortar 0.291 W/m K (recycled material from concrete demolition). Rico Rodríguez et al. [61] reported a value of 0.21 W/m K in the thermal conductivity of a cement composite material reinforced with bagasse ash and fiber. In this context, the thermal conductivity value of 0.2008 W/m·K of the PO + CE panel proposed in this study has a low thermal conductivity suitable for thermal applications in construction.

4. Conclusions

The bioclimatic analysis for the city of Oaxaca showed internal heating and thermal mass requirements for thermal comfort conditions. A hybrid panel composed of cane bagasse ash, corncob, and lime was proposed to reach such comfort conditions. A sandwich-type construction component was made with the hybrid panel and two outer layers of reinforced mortar (ferrocement) and the middle of the cane bagasse ash and corncob panel (2F + POCE). Five experimental chambers were built to determine the surface thermal performance of the panels (2F + POCE), (PO + CE), (2F + A), (CLC), and (LF). The month of March was selected for the analysis based on the climatic normals of the locality. The best decrement factor and thermal lag were presented in the panel consisting of two ferrocement slabs and the waffle panel (2F + POCE), registering 0.19 and 6 hrs, respectively. These results are significant because the panel limits the heat flow at peak hours when temperatures reach their highest values. Therefore, such a composite can provide an environmentally friendly alternative for energy savings and thermal comfort.

Acknowledgements

The authors express their sincere thanks to the National Council for Science and Technology (CONACyT), Secretary of Public Education (SEP), National Polytechnic Institute IPN/CIIDIR Oaxaca, and National Technological Institute of Mexico/Oaxaca Institute of Technology for the financial support. This work was developed within the research framework of the SIP project 20195480.

Conflict of interest

The authors declare that they have no known competing financial interests or personal relationships that could have appeared to influence the work reported in this work.

Author details

Rafael Alavéz-Ramírez[1], Fernando Chiñas-Castillo[2]*, Magdaleno Caballero-Caballero[1], Valentín Juventino Morales-Domínguez[1], Margarito Ortiz-Guzmán[1], Maria Eugenia Silva-Rivera[1], Roberto Candido Jimenez-Piñon[2] and Angel Ramos-Alonso[2]

1 National Polytechnic Institute IPN, CIIDIR Oaxaca, Santa Cruz Xoxocotlán, Oaxaca, México

2 Department of Mechanical Engineering, National Technological Institute of Mexico/ Oaxaca Institute of Technology, Oaxaca de Juarez, Oax, México

*Address all correspondence to: fernandochinas@gmail.com

References

[1] Manzano-Agugliaro F, Montoya FG, Sabio-Ortega A, García-Cruz A. Review of bioclimatic architecture strategies for achieving thermal comfort. Renewable and Sustainable Energy Reviews. 2015;**49**: 736-755. DOI: 10.1016/j.rser.2015.04.095

[2] Wegertseder P, Schmidt D, Hatt T, Saelzer G, Hempel R. Barreras y oportunidades observadas en la incorporación de estándares de alta eficiencia energética en la vivienda social chilena. Arquit y Urban. 2014;**XXXV**: 37-49

[3] Ofori G, Briffett C IV, Gang G, Ranasinghe M. Impact of ISO 14000 on construction enterprises in Singapore. Construction Management and Economics. 2000;**18**:935-947. DOI: 10.1080/014461900446894

[4] Gino Moncada LG, Asdrubali F, Rotili A. Influence of new factors on global energy prospects in the medium term: Comparison among the 2010, 2011 and 2012 editions of the IEA's world energy outlook reports. Economics and Policy of Energy Environment. 2013;**3**: 67-89. DOI: 10.3280/EFE2013-003003

[5] Zhang Y, Wang J, Hu F, Wang Y. Comparison of evaluation standards for green building in China, Britain, United States. Renewable and Sustainable Energy Reviews. 2017;**68**:262-271. DOI: 10.1016/J.RSER.2016.09.139

[6] WBCSD. Energy Efficiency in Buildings Facts and Trends: Business Realities and Opportunities. Switzerland: WBCSD. July 2008. ISBN: 978-3-940388-26-1

[7] Bressand F, Farrell D, Haas P, Morin F. Curbing Global Energy Demand Growth: The Energy Productivity Opportunity. McKinsey Glob Inst. Sao Paulo, Brazil: McKinsey&Company; 2007. pp. 1-290

[8] Secretaria del Medio Ambiente del Gobierno del Distrito Federal. Gac Of Del Dist Fed: Programa de certificación de edificaciones sustentables; 2012. pp. 1-90

[9] Mounir S, Khabbazi A, Khaldoun A, Maaloufa Y, El Hamdouni Y. Thermal inertia and thermal properties of the composite material clay-wool. Sustainable Cities and Society. 2015;**19**: 191-199. DOI: 10.1016/j.scs.2015.07.018

[10] Wang L, Zhou Q, Ji X, Peng J, Nawaz H, Xia G, et al. Fabrication and characterization of transparent and uniform cellulose/polyethylene composite films from used disposable paper cups by the "one-pot method". Polymers (Basel). 2022;**14**:1070-1085. DOI: 10.3390/polym14061070

[11] Soret GM, Vacca P, Tignard J, Hidalgo JP, Maluk C, Aitchison M, et al. Thermal inertia as an integrative parameter for building performance. Journal of Building Engineering. 2021;**33**: 101623. DOI: 10.1016/j.jobe.2020.101623

[12] Chikhi M. Young's modulus and thermophysical performances of bio-sourced materials based on date palm fibers. Energy and Buildings. 2016;**129**: 589-597. DOI: 10.1016/j.enbuild.2016.08.034

[13] Çomak B, Bideci A, Salli BÖ. Effects of hemp fibers on characteristics of cement based mortar. Construction and Building Materials. 2018;**169**:794-799. DOI: 10.1016/j.conbuildmat.2018.03.029

[14] Hamza S, Saad H, Charrier B, Ayed N, Charrier-El BF. Physico-chemical characterization of Tunisian

plant fibers and its utilization as reinforcement for plaster based composites. Industrial Crops and Products. 2013;**49**:357-365. DOI: 10.1016/j.indcrop.2013.04.052

[15] Cherki AB, Remy B, Khabbazi A, Jannot Y, Baillis D. Experimental thermal properties characterization of insulating cork-gypsum composite. Construction and Building Materials. 2014;**54**:202-209. DOI: 10.1016/j.conbuildmat.2013.12.076

[16] Khalil A, Khalil HPSA, Alwani MS, Mohd Omar AK. Chemical composition, anatomy, lignin distribution, and cell wall structure of Malaysian plant waste fibers. BioResources. 2006;**1**:220-232

[17] Loh YR, Sujan D, Rahman ME, Das CA. Sugarcane bagasse—The future composite material: A literature review. Resources, Conservation and Recycling. 2013;**75**:14-22. DOI: 10.1016/J.RESCONREC.2013.03.002

[18] Manohar K. Experimental investigation of building thermal insulation from agricultural by-products. British Journal of Applied Science and Technology. 2012;**2**:227-239. DOI: 10.9734/bjast/2012/1528

[19] Aminudin E, Khalid NHA, Azman NA, Bakri K, Din MFM, Zakaria R, et al. Utilization of Baggase waste based materials as improvement for thermal insulation of cement brick. MATEC Web Conf. 2017;**103**: 01019. DOI: 10.1051/matecconf/2017 10301019

[20] Canilha L, Chandel AK, Suzane Dos Santos Milessi T, Antunes FAF, Luiz Da Costa Freitas W, Das Graças Almeida Felipe M, et al. Bioconversion of sugarcane biomass into ethanol: An overview about composition, pretreatment methods, detoxification of hydrolysates, enzymatic saccharification, and ethanol fermentation. Journal of Biomedicine & Biotechnology. 2012;**2012**: 2012. DOI: 10.1155/2012/989572

[21] Watt JD, Thorne DJ. Composition and pozzolanic properties of pulverised fuel ashes.* I. composition of fly ashes from some component particles British power stations and properties of their component particles. Journal of Applied Chemistry. 1965;**15**:585-594

[22] Moraes JCB, Melges JLP, Akasaki JL, Tashima MM, Soriano L, Monzó J, et al. Pozzolanic reactivity studies on a biomass-derived waste from sugar cane production: Sugar cane straw ash (SCSA). ACS Sustainable Chemistry & Engineering. 2016;**4**:4273-4279. DOI: 10.1021/acssuschemeng.6b00770

[23] Payá J, Monzó J, Borrachero MV, Díaz-Pinzón L, Ordónez LM. Sugar-cane bagasse ash (SCBA): Studies on its properties for reusing in concrete production. Journal of Chemical Technology and Biotechnology. 2002;**77**: 321-325. DOI: 10.1002/jctb.549

[24] Neville AM. Tecnología del concreto: Curado del concreto. Mexico D.F: Instituto Mexicano del Cemento y del Concreto, A. C; 2013

[25] Martirena Hernandez JF, Middendorf B, Gehrke M, Budelmannt H. Use of wastes of the sugar industry as pozzolana in lime-pozzolana binders: Study of the reaction. Cement and Concrete Research. 1998;**28**: 1525-1536. DOI: 10.1016/S0008-8846 (98)00130-6

[26] Cordeiro GC, Toledo Filho RD, Tavares LM, Fairbairn EMR. Pozzolanic activity and filler effect of sugar cane bagasse ash in Portland cement and lime mortars. Cement and Concrete Composites. 2008;**30**:410-418.

DOI: 10.1016/J.CEMCONCOMP.2008.01.001

[27] Cordeiro GC, Toledo Filho RD, Fairbairn EMR. Effect of calcination temperature on the pozzolanic activity of sugar cane bagasse ash. Construction and Building Materials. 2009;**23**:3301-3303. DOI: 10.1016/J.CONBUILDMAT.2009.02.013

[28] Ganesan K, Rajagopal K, Thangavel K. Evaluation of bagasse ash as supplementary cementitious material. Cement and Concrete Composites. 2007;**29**:515-524. DOI: 10.1016/J.CEMCONCOMP.2007.03.001

[29] Jauberthie R, Rendell F, Tamba S, Cisse I. Origin of the pozzolanic effect of rice husks. Construction and Building Materials. 2000;**14**:419-423. DOI: 10.1016/S0950-0618(00)00045-3

[30] Chandrasekhar S, Satyanarayana KG, Pramada PN, Raghavan P, Gupta TN. Processing, properties and applications of reactive silica from rice husk-an overview. Journal of Materials Science. 2003;**38**:3159-3168

[31] Chandrasekhar S, Pramada PN, Majeed J. Effect of calcination temperature and heating rate on the optical properties and reactivity of rice husk ash. Journal of Materials Science. 2006;**41**:7926-7933. DOI: 10.1007/s10853-006-0859-0

[32] Frías M, Villar-Cociña E, De Rojas MIS, Valencia-Morales E. The effect that different pozzolanic activity methods has on the kinetic constants of the pozzolanic reaction in sugar cane straw-clay ash/lime systems: Application of a kinetic–diffusive model. Cement and Concrete Research. 2005;**35**:2137-2142. DOI: 10.1016/J.CEMCONRES.2005.07.005

[33] Wu N, Ji T, Huang P, Fu T, Zheng X, Xu Q. Use of sugar cane bagasse ash in ultra-high performance concrete (UHPC) as cement replacement. Construction and Building Materials. 2022;**317**:125881. DOI: 10.1016/J.CONBUILDMAT.2021.125881

[34] Yadav AL, Sairam V, Srinivasan K, Muruganandam L. Synthesis and characterization of geopolymer from metakaolin and sugarcane bagasse ash. Construction and Building Materials. 2020;**258**:119231. DOI: 10.1016/J.CONBUILDMAT.2020.119231

[35] Mehrzad S, Taban E, Soltani P, Samaei SE, Khavanin A. Sugarcane bagasse waste fibers as novel thermal insulation and sound-absorbing materials for application in sustainable buildings. Building and Environment. 2022;**211**:108753. DOI: 10.1016/J.BUILDENV.2022.108753

[36] Brito FMS, Bortoletto Júnior G, Paes JB, Belini UL, Tomazello-Filho M. Technological characterization of particleboards made with sugarcane bagasse and bamboo culm particles. Construction and Building Materials. 2020;**262**:120501. DOI: 10.1016/J.CONBUILDMAT.2020.120501

[37] Gharieb M, Rashad AM. An initial study of using sugar-beet waste as a cementitious material. Construction and Building Materials. 2020;**250**:118843. DOI: 10.1016/J.CONBUILDMAT.2020.118843

[38] Jagadesh P, Ramachandra Murthy A, Murugesan R. Effect of processed sugar cane bagasse ash on mechanical and fracture properties of blended mortar. Construction and Building Materials. 2020;**262**:120846. DOI: 10.1016/J.CONBUILDMAT.2020.120846

[39] Jittin V, Bahurudeen A. Evaluation of rheological and durability

characteristics of sugarcane bagasse ash and rice husk ash based binary and ternary cementitious system. Construction and Building Materials. 2022;**317**:125965. DOI: 10.1016/J. CONBUILDMAT.2021.125965

[40] Tripathy A, Acharya PK. Characterization of bagasse ash and its sustainable use in concrete as a supplementary binder – A review. Construction and Building Materials. 2022;**322**:126391. DOI: 10.1016/J. CONBUILDMAT.2022.126391

[41] Souza NSL de, Anjos MAS dos, Sá M, de Das VVA, de Farias EC, de Souza MM, Branco FG, et al. Evaluation of sugarcane bagasse ash for lightweight aggregates production. Construction and Building Materials 2021;271:121604. DOI: 10.1016/J.CONBUILDMAT.2020. 121604

[42] Subedi S, Arce GA, Hassan MM, Barbato M, Mohammad LN, Rupnow T. Feasibility of ECC with high contents of post-processed bagasse ash as partial cement replacement. Construction and Building Materials. 2022;**319**:126023. DOI: 10.1016/J.CONBUILDMAT. 2021.126023

[43] Klathae T, Tanawuttiphong N, Tangchirapat W, Chindaprasirt P, Sukontasukkul P, Jaturapitakkul C. Heat evolution, strengths, and drying shrinkage of concrete containing high volume ground bagasse ash with different LOIs. Construction and Building Materials. 2020;**258**:119443. DOI: 10.1016/J.CONBUILDMAT. 2020.119443

[44] Klathae T, Tran TNH, Men S, Jaturapitakkul C, Tangchirapat W. Strength, chloride resistance, and water permeability of high volume sugarcane bagasse ash high strength concrete incorporating limestone powder. Construction and Building Materials. 2021;**311**:125326. DOI: 10.1016/J. CONBUILDMAT.2021.125326

[45] Barbosa FL, Cordeiro GC. Partial cement replacement by different sugar cane bagasse ashes: Hydration-related properties, compressive strength and autogenous shrinkage. Construction and Building Materials. 2021;**272**:121625. DOI: 10.1016/J.CONBUILDMAT.2020. 121625

[46] Athira G, Bahurudeen A. Rheological properties of cement paste blended with sugarcane bagasse ash and rice straw ash. Construction and Building Materials. 2022;**332**:127377. DOI: 10.1016/J.CONBUILDMAT.2022. 127377

[47] Fuentes FV. Mapas bioclimáticos de la República Mexicana. 1a. ed. México D. F: Universidad Autónoma Metropolitana; 2014

[48] Köppen WP. Climatologia, con un estudio de los climas de la tierra. Mexico D.F: Mexico, Fondo de Cultura Economica; 1948

[49] Garcia E. Modificaciones al sistema de clasificacion climatica de Koppen. 2004th ed. Mexico D.F: Instituto de Geografia UNAM; 2004

[50] Carvalho W, Canilha L, Castro PF, Barbosa LDFO. Chemical composition of the sugarcane bagasse. In: Society for Industrial Microbiology, Editor. 31st Symp. Biotechnol. Fuels Chem. Society for Industrial Microbiology: San Francisco CA; 2015

[51] Luz SM, Gonçalves AR, Ferrão PMC, Freitas MJM, Leão AL, Del Arco AP Jr. Water absorption studies of vegetable fibers reinforced polypropylene composites. In: Proceeding of 6th Int. Symp. Nat. Polym. Compos. ISNaPol 6

and XI International Macromolecular Colloquium IMC 11. Gramado, RS Brazil: Associacao Brasileira de Polimeros (ABPol); 22-25 April, 2007. pp. 73-78

[52] Ramlee NA, Naveen J, Jawaid M. Potential of oil palm empty fruit bunch (OPEFB) and sugarcane bagasse fibers for thermal insulation application – A review. Construction and Building Materials. 2021; **271**:121519. DOI: 10.1016/J.CONBUILD MAT.2020.121519

[53] Alavéz-Ramírez R, Chiñas-Castillo F, Morales-Domínguez VJ, Ortiz-Guzmán M, Caballero-Montes JL, Caballero-Caballero M. Thermal lag and decrement factor of constructive component reinforced mortar channels filled with soil–cement–sawdust. Indoor and Built Environment. 2018;**27**: 466-485. DOI: 10.1177/ 1420326X16676611

[54] Alavéz Ramírez R, Morales Domínguez VJ, Ortiz GM. Mortero a base de Olote, ceniza de bagazo de caña y cal como relleno ligero. 6to. Congr. Int. Virtual Innovación Tecnológica y Educ. Ediciones ILCSA S.A. de C.V: CIVITEC, Tijuana, Baja California, México; 2018

[55] Alavez-Ramirez R, Chiñas-Castillo F, Morales-Dominguez VJ, Ortiz-Guzman M. Thermal conductivity of coconut fibre filled ferrocement sandwich panels. Construction and Building Materials. 2012;**37**:425-431. DOI: 10.1016/j. conbuildmat.2012.07.053

[56] Belhadj B, Bederina M, Makhloufi Z, Goullieux A, Quéneudec M. Study of the thermal performances of an exterior wall of barley straw sand concrete in an arid environment. Energy and Buildings. 2015;**87**:166-175. DOI: 10.1016/j. enbuild.2014.11.034

[57] Roux-Gutiérrez RS, Velázquez LJ. Bloques de Tierra Comprimida, Su Retardo Térmico e Impacto Ambiental. Rev Legado Arquit y Diseño. Mexico: Universidad Autónoma del Estado de Mexico; 2016. pp. 1-13. ISSN: 2448-749X

[58] Gallegos-Ortega R, Magaña-Guzmán T, Reyes-López JA, Romero-Hernández MS. Thermal behavior of a straw bale building from data obtained in situ. A case in northwestern México. Building and Environment. 2017;**124**: 336-341. DOI: 10.1016/j. buildenv.2017.08.015

[59] Ruiz Torres RP. Evaluación del Sistema Termolosa entre la medición experimental y el calculado con la NMX-C-460-ONNCCE-2009. Vivienda y Comunidades Sustentables. 2019;**2019**: 119-136. DOI: 10.32870/rvcs.v0i6.126

[60] Borbón Almada AC, Alpuche Cruz MG, Miranda Pasos I, Marincic Lovriha I, Ochoa de la Torre JM. Materiales reciclados aligerados y su influencia en el consumo de energía eléctrica en viviendas económicas. Acta Univ. 2019;**29**:1-15. DOI: 10.15174/ au.2019.2096

[61] Rico Rodríguez I, Vargas Galarza Z, García Hernández E, Salgado Delgado R, Cárdenas Valdez RC, Olarte PA. Evaluación térmica de material compuesto de cemento portland reforzado con agregado fino de CBC y FO tratada con Silano. Ing Investig y Tecnol. 2020;**21**:1-11. DOI: 10.22201/ fi.25940732e.2020.21n1.001

Chapter 4

Economic Importance and Yield Potential of Sugarcane in Pakistan

Shahid Afghan, Muhammad Ehsan Khan, Waqas Raza Arshad, Karim Bukhsh Malik and Amin Nikpay

Abstract

Sugarcane is mainly cultivated in the tropical and subtropical regions of the world, and nearly 85% of sugar is used worldwide. The area, production, and yield of sugarcane has been increased worldwide as well as in Pakistan as compared to other crops. It is the second largest economically important crop after cotton. It is a high-value cash crop that has significance for sugar industries in Pakistan. It contributes about 0.6% to the GDP and 2.9% of the total value added in agriculture. It creates huge revenue for the government and is used as a source of energy/power. The climate of Pakistan is favorable for sugarcane production in Punjab and Sindh provinces. Different climatic factors, i.e., sunlight, temperature, germination, tillering, growth, humidity, dew, frost, hailstorm, windstorm, sunburn, and drought, significantly affect the production of sugarcane. Pakistan is a principal cane-growing country and stands at the fifth position in the area, sixth position in cane sugar production, and ninth largest sugar producer in the world. This chapter describes the economic importance, climate, and yield potential of sugarcane in Pakistan.

Keywords: economic significance, yield potential, sugarcane, Pakistan

1. Introduction

Sugarcane is a globally significant crop as it provides nearly 85% of the sugar consumed worldwide. In recent years, the planting of sugarcane for the production of biofuels has expanded rapidly as cases of energy canes. The cultivation of sugarcane is one of the most important activities around the world due to their alimentary, environmental, social and economic implications, and potential productive diversification with coproducts and byproducts [1]. Sugarcane is a worldwide crop cultivated in more than 105 countries. From its very origin, in earlier times to its present-day production, sugarcane has played its role in improving the socioeconomic conditions of human society [2].

Previous trends have shown a tremendous increase in area under sugarcane production throughout the world. However, for 2019–2020, global sugar production is estimated to decline 3% to 1745 million tons while sugarcane was grown on 26.5 million ha. This diminution is attributed to the 5 million tons drop in India's production, resulting from lower area and expected yields. However, Brazil and India are essentially tied as top sugar-producing countries, while Pakistan occupies the fifth place [3].

It is mostly used as a food crop for the production of raw and refined sugar, *gur*, and *shakkar*. Sugarcane improvement has traditionally focused on sucrose-yield traits. In the future, energy canes with higher yields of fermentable sugars and fiber (bagasse) for biofuel and electricity applications will be developed [4]. Sugarcane has gained importance for its dietary value and its industrial utilization for several products. Its products and byproducts have revolutionized a native and international trade, and the crop production trends have played a dominant role in altering the economic and fiscal position of countries.

Its juice is used for making white sugar, brown sugar (*khand*), and jaggery (*gur*) [5, 6]. The sugarcane bagasse, a byproduct, is nowadays valued by producers from the sugar sector since it is presented as the main feedstock source for bioenergy and biofuel production. In Pakistan, almost 3.50 million metric tons of bagasse is consumed with an average recovery rate of 30%. Bagasse is conventionally recycled as confined energy in sugar factories, i.e., boiler steam is required to generate and drive the prime movers [7].

Bagasse is used for the production of compressed fiberboard, paper, plastics, and furfural. Owing to the high silicon content on sugarcane, bagasse can be used to produce silica, a valued material for industry and health products. Molasses are used in distilleries to manufacture ethyl alcohol, butyl alcohol, citric acid, etc. Sugarcane filter cake (press mud) has good potential as organic fertilizer [8].

2. Importance of sugarcane

2.1 Dietary value

Sugar is used as a sweetener in many dishes of varying tastes, beverages, and pharmaceutical products. It is an important constituent of human diet, having a pleasant taste and a high calorific value of 387 per 100 grams [7]. Some fruits and vegetables also have some forms of sugar (**Table 1**), but it is not derived as a commercial product [10].

The trend of using low-calorie sweeteners (LCS) has now developed in some quarters. Among the low-energy sweeteners, saccharin is the oldest sweetener [11]. Sugarcane juice contains zero fat, cholesterol, fiber, and protein [11]. This is the healthiest and the most nutritious drink one can think of consuming [12]. The latest research showed that both sugar-sweetened beverages (SSBs) and LCS beverages were linked with an increased risk of developing type 2 diabetes. Sugar percent in different fruits has been presented in **Table 1** [12]. Some forms of sugar are produced from palm trees, sugar maple, sweet sorghum, maize, sugar beet, and sugarcane (**Table 2**) [2].

2.2 Grower's prosperity

Sugarcane is a high-value cash crop that has significance for sugar and sugar-related industries in Pakistan. It contributes about 0.6% to the GDP and 2.9% of the total value added in agriculture and has brought prosperity to grower's community [14]. In areas of cane concentration, healthy socioeconomic change is witnessed, which has improved the living standard of growers. The dilemma is that farmers normally do not receive a fair market price for their sugar crop [15].

General problems faced by almost every farmer regarding sugarcane production are lack of irrigation water, nonavailability of improved varieties of sugarcane, high

Fruits	Sugar (%)	Fruits	Sugar (%)
Mangoes	12.7	Oranges	8
Strawberries	4.1	Apricots	7.8
Pomegranates	11.6	Plums	8.4
Guavas	4.7	Pears	8.4
Papaya	5	Pineapple	8.4
Watermelon	5.3	Cherries	10.9
Grapefruit	5.9	Apples	8.8
Figs	13.8	Grapes	13.8
Peaches	7.1	Bananas	10.4

Source: [9].

Table 1.
Sugar contents in different fruits.

Palm trees	9% sucrose on the weight of palm juice
Sugar maple	2% sugar maple sap
Sweet sorghum	9% sucrose (brix 19%, pol 15%).
Maize	2–4% glucose
Sugar beet	12–18% sucrose
Sugarcane	10–13% sucrose

Source: [13].

Table 2.
Plants utilized for commercial sugar production.

cost of inputs for land preparation, diseases, insect pest, weeds, and marketing problems [16]. The adversities faced by sugarcane farmers have been unequivocally portrayed by the multitude of protests organized by sugarcane farmers. Given the failure of the governments to resolve these issues, it is apparent that a different course of action needs to be taken to address the fundamental problem. We need to look beyond our conventional strategies.

In developing countries, the sugar factories delay the timely payments of sugarcane farmers. In view of millions of sugarcane farmers, the governments need to swiftly resolve the most pivotal issue before it further retrogrades. In return, governments mainly have two options. First, one is to pay back the dues to the farmers from the state treasury or provide subsidies to sugar factories. Either way, the government will have to shift the monetary burden to the taxpayers [5]. In the short run, given the obstinate demand of sugarcane farmers to receive their payment and the incapability of sugar factories to make the payments, this seems to be a plausible option.

2.3 Sugar industry of Pakistan

A total of 89 sugar mills are operational [17]. Sugarcane cultivated on an area of 1.16 million ha during 2020–2021 [3]. On regional basis, Punjab's share in the total sugarcane

Province	Area (1000 ha)	Share in cropped area (%)	Prod. (1000 ton)	Provincial share in prod. %	Yield (t/ha)
Punjab	643.4	4.71	43,346.6	65.7	67.37
Sindh	286.0	9.95	17,233.8	26.8	60.24
KPK	109.4	6.31	5753.9	7.5	52.62
Pakistan	1038.9	5.37	66,334.4	100.0	63.85

Source: [3].

Table 3.
Provincial distribution of sugarcane production in Pakistan in 2020.

area of Pakistan was 65.7% followed by Sindh 26.8% and Khyber Pakhtunkhwa (KPK) was 7.5%. The yield of sugarcane was increased from last five in 2020–2021; the highest yield was recorded 69.55 t/ha, while the maximum share in yield was of Punjab at 59.5 tons (**Table 3**).

The area under sugarcane in Pakistan has increased from 960,000 has in 2001–2002 to 1,040,000 ha in 2019–2020 with an annual growth rate of 1.0% [18]. The total increase in area, production, and per hectare yield during this period is around 26.7, 73.1, and 27.5% during the period [3].

2.4 Labor and workforce

Sugarcane is a source of employment for millions of people both at the farm sector and in sugar industry [19]. For example, in Pakistan, all most 15,000–25,000 families are involved in the sugarcane zone of one factory. Almost 980,000 farmers were engaged with sugarcane cultivation in Pakistan, out of which 707,000 farmers worked in Punjab, 200,000 in Sindh, and 81,000 in KPK. On average, 1200 employees working in a sugar mill with workforce of 106,800 individuals in the production and processing departments. Overall, around 4 million employees are engaged directly or indirectly in sugar business in Pakistan [20].

Owing to the lack of mechanical sugarcane farming (planting and harvesting) in Pakistan, innumerable workers are deployed in cane harvesting, transport, and loading and unloading of cane. The huge involvement of technicians in farm machinery with the transport system of tractors, trolleys, and trucks depicts the quantum of business in various sectors [5].

2.5 Source of food fodder and fertilizer

Sugarcane is used as a source of food and fodder. The press mud and fly ash products of sugar industry are high-valued organic fertilizers. Filter cake is used as fertilizer and used in brick kilns [21] also used with distillery effluent and nitrogen-fixing bacteria. Some processors/factories have acquired certificates of halal and organic products like sugar/*gur*/*shakkar* [6]. The production of organic *gur* and *shakkar* is about 1000 tons having a business of around 300 million [6, 22].

2.6 Government revenues

In Pakistan, the revenue is derived in the form of general sale tax (GST) on the sale of sugar, which is 17% of the sugar price. A substantial amount is recovered as sugarcane Cess Fund @ Rs. 2.00 per maund of sugarcane, delivered to sugar mills. Land revenue and water rates, @ Rs. 400 per acre of cane, amount to billions of rupees per annum and is the direct source of income for the government. The excise duty on molasses and alcohol is also a big source of revenue for the government as well [23].

2.7 Source of energy/power

Molasses, the main byproduct of the sugar industry, is utilized for the production of alcohol, rum, and as feed for livestock. Ethanol has now gained high importance as an energy product [24]. Sugarcane is used as feedstock for ethanol production and as biofuel in vehicles. The use of high-pressure steam boilers and efficient use of energy in the sugar factories has given new vistas of cogeneration from bagasse and trash to export surplus electric power to the national grid system [25]. The production of biogas from spent wash/vinasse fermentation is also a source of cheaper energy in distilleries [24, 25].

2.8 Industrial utilization

Sugarcane occupies a prominent place for high biomass-fiber production and molasses. As a raw material, sugarcane has attained worldwide importance for dozens of industrial derivatives. The main byproducts of cane are molasses used for alcohol, bagasse utilized for energy production, cogeneration, the manufacture of particle-board, and filter press cake used as organic fertilizer [26].

In the present-day market economy, the profits derived from the manufacture of sugar are getting low, and more attention is paid to the manufacture of byproducts. Today, the focus is to adopt sugar production technology to yield energy-based coproducts, ethanol, and electricity. Cane acreage fluctuates depending upon market forces, socioeconomic conditions, and crop competitions [24].

3. Cane yields on the global level

On the global level, sugarcane is grown in 105 countries located in equatorial, tropical, and subtropical regions. Cane area, production, and yields of principal cane-growing countries are shown in **Table 4**. The world data show that during 2020, sugarcane was grown on an area of 25.98 million hectares with a cane yield of 70.9 tons per hectare (t/ha). Concerning area, Brazil occupies the leading position by growing cane in an area of 10.18 million hectares followed by India, China, Thailand, and Pakistan growing cane in an area of 4.39, 1.38, 1.37, and 1.22 million hectares, respectively [28].

During the midst of the twentieth century world, the average yield was hardly 42.5 t/ha, which has now increased to 70.9 t/ha. During 1950–1953 period, principal cane-growing countries, Brazil, India, China, and Thailand, had the cane yields of 38.7, 32.1, 35.2, and 17.5 t/ha, respectively, much below the world average. During five decades of progressive development, Brazil, China, and Thailand improved their yields to a level of 74.5, 76.1, and 75.2 t/ha, respectively, which exceeds the world average of 70.9 t/ha. Still, there are countries with an average yield of 121.0 tons per hectare (Guatemala) and 112.7 tons per hectare (Egypt) as shown in **Table 5** [30].

Pakistan grasps an important position in cane area and production in the world and ranks on top fifth position, but with respect to cane yield (69.55 t/ha), it ranks

Year	Punjab		Sindh		KPK		Pakistan	
	Area (ha)	Yield (t/ha)	Area (ha)	Yield (t/ha)	Area (ha)	Yield (t/ha)	Area (ha)	Yield (t/ha)
1947–1950	170.0	37.7	8.10	34.1	41.30	28.40	201.9	33.47
1959–1960	286.6	27.09	20.20	41.70	65.20	30.40	396.6	26.80
1969–1970	454.9	43.3	83.40	40.9	81.70	39.7	620.00	42.50
1979–1980	501.4	38.7	129.90	35.90	87.10	39.20	718.50	38.3
1989–1990	501.0	37.30	250.7	49.30	102.10	43.40	854.3	41.5
1999–2000	672.1	40.3	230.6	51.27	108.3	46.3	1009.8	45.9
2009–2010	607.4	51.6	233.9	57.7	100.8	44.7	942.9	52.36
2016–2017	777.82	63.78	320.51	63.05	118.57	47.50	12.16.9	62.00
2017–2018	859.13	64.10	333.30	61.85	148.53	51.2	1340.9	62.11
2018–2019	710.61	63.19	279.50	59.72	110.10	48.8	1101.1	60.97
2019–2020	643.430	67.37	286.090	62.93	109.36	52.62	1038.9	63.85
2020–2021	776.98	73.36	279.69	65.56	107.44	52.38	1164.12	69.55

Source: [27].

Table 4.
Growth in area and yields of cane in cane-growing provinces of Pakistan from 1947 to 1950 to 2020–2021.

Country	Area (1000 ha)	Production (million/t)	Yield (t/ha)
World	25,976.94		70.89
Brazil	10,184.34	758.55	74.48
India	4389.00	306.07	69.74
China	1377.11	104.79	76.10
Thailand	1368.27	102.95	75.24
Pakistan	1216.89	73.40	60.31
Mexico	722.00	56.95	73.78
Australia	457.47	36.56	80.62
Philippines	439.47	29.29	66.94
Indonesia	430.11	21.21	49.32
Colombia	397.39	34.64	87.16
Cuba	387.70	16.07	47.45

Country	Area (1000 ha)	Production (million/t)	Yield (t/ha)
Argentina	378.82	19.16	50.54
USA	365.88	30.15	82.41
Vietnam	281.15	18.36	65.29
Guatemala	278.97	33.76	121.01
South Africa	264.50	17.39	65.74
Burma	163.25	10.37	63.52
Bolivia	151.99	8.05	52.96
Cameroon	136.29	1.29	9.45
Egypt	135.40	15.26	112.70

Source: [29].

Table 5.
Sugarcane area, production, and yields of top 20 cane-growing countries during 2017.

much below the principal cane-growing countries [31]. Compared with other countries, it appears that Pakistan took a late start to meet yield gaps. Even India has touched 70 t/ha, while Pakistan has yet to travel a lot to reach this yield level.

The highest cane yield of 121.01 t/ha is reported in Guatemala, followed by 112.70 t/ha in Egypt. Among the top 20 cane-growing countries are Colombia (87.16 t/ha), the USA (82.41 t/ha), and Australia (80.62 t/ha), while China, Thailand, Brazil, and Mexico have exceeded 70 tons in yield [28]. Pakistan has just 60 tons against about 70 tons yield of India. Though Pakistan stands at the fifth position in cane area and production in the world it has to travel a lot in the compatible field of yields per hectare.

4. Effect of latitude

The data in **Table 6** indicate that world's largest cane area is located in South America (44.5%), closely followed by Asia (40.3%). The continents of North America, Africa, and Oceania grow cane in a small fraction of 8.85, 8.91, and 1.82%

Continent	Latitude ranges—cane area and (cane yield)					% Share in area
	0–10	10–20'	20–25'	25–35'	Total	
Asia						
Area	691.8	3496.4	2421.4	3869.3	10,478.9	40.3
Yield	61.1	66.3	66.5	62.5	64.7	
Africa						
Area	523.5	400.9	138.9	468.3	1531.6	5.9
Yield	34.5	52.6	42.7	82.8	54.7	
North America						
Area	765.0	14.8	1162.0	365.9	2307.7	8.9
Yield	89.2	59.7	62.9	82.4	74.7	
South America						
Area	733.0	10,336	118.0	385.4	11,572.7	44.5
Yield	82.7	74.2	56.0	50.7	74.6	

Continent	Latitude ranges—cane area and (cane yield)					% Share in area
	0–10	10–20'	20–25'	25–35'	Total	
Europe						
Area	—	—	—	0.08	0.08	0.00
Yield				27.0	27.0	
Oceania						
Area	7.1	35.2	435.5	—	477.8	1.8
Yield	28.7	42.3	80.6		73.4	
Total world						
Area	2720.4	14,283.6	4275.8	5088.9	25,976.9	100
Yield	70.0	71.7	65.9	64.9	70.9	
% Share in area	10.5	55.0	16.5	19.6	100	

Source: [32].

Table 6.
Cane area (thousand ha) and cane yield (t/ha) of various continents located in different latitude ranges, during 2017.

area, respectively. The maximum cane area (55%) falls in countries situated in latitude ranges of 10–20° N or S. The countries in latitude ranges of 0–10°, 20–25°, and 25–35° grow 10.5, 16.5, and 19.5% of the total world cane average, respectively. The continent of Asia has a relatively higher proportion of cane grown in 25–35° latitudes, South America in 10–20°, while cane area in North America and Oceania mostly fall in latitude ranges of 20–25°.

The latitudes have a great impact on cane yields. The highest cane yields are observed in Latitude ranges of 0°–20° (70.0–71.7 t/ha) followed by 20°–25° (65.9 t/ha) and 25°–35° (64.9 t/ha). It indicates that as we go away from the equator, there is a gradual decline in cane yields. With respect to overall global cane yields, North America and South America have the highest yield of 74.7 t/ha and is closely followed by Oceania (73.4 t/ha), while the yields of Asia (64.7 t/ha), Africa (54.7 t/ha), and Europe (27.0 t/ha) are considerably low. The latitude range of 10°–25° is considered the most favorable for cane production.

5. Correlation of latitude with cane yields and juice quality

5.1 Cane yield

On the global level, the effect of latitude on cane yields is well marked [33]. The yield level toward the equator is high, and as the cane growing is shifted away from the equator, yields show a gradual decline [34]. The survey of the global area indicates some very interesting climatic features of countries lying in favorable climatic zones but having very low yields [34] (**Table 7**).

In present-day agriculture, the word "favorable climate" has been replaced by "favorable environment" [8, 34]. High yields of cane and sugar are obtained under such dry environments where satisfactory soil water balance can be achieved by irrigation water under optimum soil and crop management practices [12, 13]. The sugarcane varieties have been developed, which owing to their great adaptability can be cultivated under a wide range of climate and soil conditions [16]. The cultivation of

District	Cane area (thousand ha)	Cane yield (t/ha)	District	Cane area (thousand ha)	Cane yield (t/ha)
Punjab					
Punjab	859.3	64.10			
Sargodha	65.96	57.18	Sheikhupura	2.02	59.30
Khushab	8.90	56.91	Nankana sb.	13.76	56.26
Mianwal	3.24	50.45	Lahore	0.40	50.17
Bhakkar	29.95	65.21	Kasur	30.76	51.37
Faisalabad	112.50	57.00	Multan	6.47	52.85
T. T. Singh	38.44	62.35	Lodhran	5.26	69.27
Jhang	55.04	59.30	Khanewal	6.88	63.36
Chiniot	43.71	55.43	Vehari	17.81	57.46
Gujrat	2.43	48.61	Muzaffargarh	61.11	68.44
M.Bahauddin	23.88	52.48	Layyah	18.21	65.58
Sialkot	1.21	35.05	D.G. Khan	10.12	69.36
Narowal	1.62	31.91	Rajanpur	32.37	82.09
Gujranwala	202	39.06	Bahawalpur	30.35	66.41
Hafizabad	6.47	48.42	R.Y. Khan	193.03	77.75
Okara	14.57	49.90	Bahawalnagar	13.76	58.11
Sahiwal	4.86	54.05			
Pakpattan	2.02	53.95			
Sindh					
Sindh	327.29	62.96			
Thatta	36.37	59.30	Nawabshah	35.71	69.19
Badin	42.60	46.45	Nawshehra Feroz	22.26	63.26
Hyderabad	6.39	57.33	Sanghar	17.34	67.21
T.M. Khan	18.88	64.14	Sukkar	7.86	66.22
T.Allahyar	20.45	45.47	Khairpur	22.76	62.27
Matiari	15.84	66.22	Ghotki	53.40	82.71
Jamshoro	0.58	44.48	Larkana	0.71	69.68
Dadu	4.50	53.37	Kambar	0.19	54.36
Mirpur Khas	18.90	55.37	Shikarpur	0.47	45.46
Tharparkar	0.28	60.29	Jaccobabad	0.15	40.52
40.52Umarkot	1.64	63.60			
Khyber Pakhtunkhwa					
Khyber Pakhtunkhwa	148.53	51.23			
Bannu	0.43	39.74	Naushehra	2.04	50.77
Charsada	32.20	43.48	Peshawar	10.71	51.94
D. I. Khan	63.08	62.00	Swabi	2.21	38.75

District	Cane area (thousand ha)	Cane yield (t/ha)	District	Cane area (thousand ha)	Cane yield (t/ha)
Malakand	4.88	38.17	Tank	1.09	39.96
Mardan	30.39	42.48	Others	1.50	25.38

Table 7.
Sugarcane area and yields of various districts in Punjab, Sindh, and Khyber Pakhtunkhwa, during 2017–2018.

Years	Sugar recovery percent cane			
	Sindh	Punjab	KPK	Pakistan
1947–1950	—	7.23	7.08	7.16
1950–1955	—	7.73	7.43	7.58
1955–1960	—	7.54	8.08	7.81
1960–1965	7.97	7.91	8.03	7.97
1965–1970	8.85	8.41	8.62	8.63
1970–1975	8.88	8.67	7.68	8.41
1975–1980	9.24	8.42	8.01	8.56
1980–1985	9.10	8.28	8.39	8.59
1985–1990	9.09	8.23	8.85	8.72
1990–1995	9.35	8.23	8.59	8.72
1995–2000	9.55	7.93	8.02	8.50
2000–2005	9.35	8.61	8.10	8.69
2005–2010	9.49	8.78	8.12	8.80
2010–2015	10.10	9.75	9.20	9.68
2015–2016	9.94	10.65	9.44	10.16
2016–2017	9.77	10.16	9.43	9.87
2017–2018	9.79	10.55	9.52	10.02
2018–2019	10.31	10.79	10.41	10.47
2019–2020	9.71	10.21	10.30	9.89
2020–2021	9.43	10.02	9.86	9.61

Source: [3].

Table 8.
Sugar recovery trends in sugar mills of Sindh, Punjab, and KPK during 1947–2020.

sugarcane is, therefore, spread in tropical, subtropical, and temperate regions between latitude ranges of 0° and 37° N and S. World data indicate that the best cane-producing areas in Asia are situated in 25°–35° latitude [35]. On a global level as well, after 10°–20° latitude, 25°–35° latitude region produces the maximum cane in the world (**Table 8**).

5.2 Cane quality

Lower latitudes assure longer days with increased sunshine duration during the growth phases of the plant. As such, high sucrose contents in cane are noted at around

18°S and 18°N. Sugar contents drop rapidly from these latitudes toward the subtropics and less rapidly toward the equator.

High sugar recoveries demand relative temperature disparity (RTD) value around 14–16°C, low daily mean temperature 10–12°C, and low relative humidity (<50%), during ripening [36]. The wider range between a day (maximum) and night (minimum) temperature during ripening results in higher sucrose contents [37]. High altitudes of 1000 mm or more within equatorial climates and low altitudes of subtropics also produce good-quality cane [38, 39]. Tropical countries or regions with high average temperature (25–27°C) and high precipitation of more than 1500 mm produce more biomass of cane, but it is of low sugar contents. In tropical areas where the weathers are distinctly wet and dry, moisture status inside the cane is a dominating factor in the synthesis and translocation of sugars. Under such conditions, control on irrigation and fertilizer is exercised to hasten or delay the maturity [40]. In modern agriculture, high sucrose content is achieved irrespective of latitude, in regions of low precipitation of both tropics and subtropics, if optimum soil and crop water balance have been achieved by human control. The climatic features have distinct behavior on plant growth and maturity phases:

- The high yield of cane per hectare is the result of a large biomass production, which is caused by moderately high temperature, high precipitation, and longer growing season. These, however, do not favor high sucrose recovery in cane [33].

- High sucrose contents in cane are a function of relatively low temperature and low precipitation during the maturity phase, which, otherwise, depress growth [36].

6. Climatic factors with effects

The change in weather factors, such as sunlight, temperature, rainfall, humidity, and solar radiation, affects the different phases of growth, maturity, and ripening phases of the plant [41].

6.1 Sunlight

Sunlight is used as photoperiod to regulate various growth processes in vegetative phase and for flowering. The sugarcane yield is governed by:

a. the amount of sunlight exposure to the crop over its lifetime, during various phases of growth

b. ability to intercept and use light energy to produce carbohydrates, the main source of sugars.

For sugarcane, the desirable locations with respect to light availability are within 30°N or °S, and more so on tropical latitude ranges. Crops having high yield and great potential need longer time span with greater light. For the best yield, 18 months are preferred as compared to 12 months [41]. Tillering is badly affected in long hazy season. For good tillering, excessive light is required; if plants grow to close in, light is curtailed that causes tiller mortality.

6.2 Temperature

Sugarcane crop might be exposed to scorching heat with a maximum mean temperature of 40–42°C reaching to the highest maximum of 48–50°C in Pakistan, India, and other countries. A little deviation was observed in maximum and minimum temperature during summer and winter months in tropical and equatorial climates, except on mountain heights or regions under the influence of cold or hot sea currents.

6.3 Germination

Germination estimates revealed that in Pakistan, the optimum temperature range was 28–33°C [42]. Germination is accelerated in the temperature range of 26–33°C, is sensitive at 22°C, and is decreased at 18°C and below. The range 32–38°C is reported to be the optimum temperature for germination.

In Pakistan, September and March considered the ideal conditions for good germination. When the temperature is above 25°C, the maximum germination of almost 75% was obtained within 20–25 days of cane planting. In February and March, temperature is somewhat lower and takes 4 or 5 weeks for complete germination. Rains may extend the germination period [43].

6.4 Tillering

In sugarcane, tillering is noticeably correlated with temperature. Tillering gradually increases with increasing temperature until the maximum is reached somewhere around 30°C.

6.5 Growth

Temperature is considered the chief growth-monitoring factor. The range 26–27°C is the ideal temperature for optimum growth; growth is checked at 10°C and 27°C is optimum for both growth and nutrient absorption [40]. The critical temperature for cane growth is 8–20°C [11, 37] and below 12°C, growth ceases; if it is less than 5°C, the leaves become pink [11, 37]. Canopy development is governed by the prevalence of moderate temperature between 21°C and 38°C with a relative humidity of 50% [44]. At a temperature of 35°C, the decrease in growth rate is due to an increase in photorespiration [4]. Cane is found to thrive at temperatures as high as 45°C in Pakistan.

The lower leaves die and the upper leaves show a yellow-green appearance. 27°C is the optimum temperature for growth and nutrient absorption [45]. A seven-fold decrease in phosphorus uptake was noticed as the temperature decreased from 22.1°C to 16.7°C [45]. A drop in root temperature from 23–19°C cuts phosphorus intake to one-third and reduces nitrogen intake to about one-half [45].

6.6 Maturity and ripening

Large differences between day and night temperatures favor the process. The range 10–12°C is considered the minimum temperature for ripening [46]. However, the ripening process is accelerated with the drop in humidity in the environment and leaf moisture in the plant [1]. To hasten the ripening process, mild drought conditions may be created by withholding irrigation.

With respect to sugar contents in cane, this process needs further clarification. Low temperature and moderate water deficit associated with nitrogen deficit are the most important ripening agents [47]. Some authors propose that in addition to air temperature and soil moisture, the variables such as photoperiod and solar radiation must also be considered [28]. Cloudy days are limiting factors for sucrose accumulation in plants. Solar radiation is directly related to sugar ripening. The factors such as air temperature, precipitation, and cloud-free bright shining days increase the time period available for photosynthesis [48]. A temperature of 20–250°C during the day and 10–14°C during the night, associated with bright sunny days with low humidity, are ideal for sucrose synthesis and accumulation in leaves.

6.7 Humidity

The growth of cane crops depends upon atmospheric humidity and proper soil moisture. High humidity for a long duration may inhibit evapotranspiration and affect growth. High humidity causes infection of some viral and fungal diseases. Evapotranspiration rate increases with high temperature and low humidity to balance the water stability in the plant, thus hindering growth, and may also increase the sensitivity of sugarcane to some sap-sacking pests such as mites. A positive correlation has been found between the rate of cane elongation and rainfall. The distribution of rainfall is more important than total rainfall, and light showers are more beneficial than heavy rains. Water needs of plants shower influence cane growth through moisture absorption by leaves, raising atmospheric humidity and keeping the leaves' surfaces clean for optimum transpiration and respiration. Moderate rainfall of 750–1000 mm supplemented by sufficient and timely irrigation is considered best for healthy crop growth. As already discussed, the ripening period demands moderate moisture stress.

6.8 Dew

Dew deposits on leaf surfaces help in foliar absorption of moisture. It also delays the rise in leaf temperature and, thus, reduces the rate of evapotranspiration. Dew deposit is estimated to be 0.25–0.40 mm per night with a total of 25–30 mm per annum. Thus, dew deposits help to mitigate the severity of water stress to a certain degree in moisture stress areas.

7. Sugarcane agro-climatic zones with yield and production potential in Pakistan

Pakistan with cane sugar production of 70 million tons during 2016–2017, ranked fifth in the world [49]. As influenced by climatic conditions and cane/sugar marketing trends, cane production varies from year to year.

The sugar industry has developed a daily cane-crushing capacity of 567,920 tons for 125-day crushing duration [19]. The Pakistan sugar industry attained a sugar recovery level of 10.47%, during 2018–2019. Considering further propagation of quality cane varieties, it is not far to achieve an average sugar recovery of 11.0% [10].

As for cane production, so far average yield is just 64.30 tons per hectare; nevertheless, some of the districts of Punjab and Sindh have achieved an average cane yield of 82 tons per hectare [14]. Improved production technology and inputs need to be further mobilized to enhance national yield to 70 tons per hectare and then to plan for 75 tons. However, sugarcane cultivation is confined in part of coastal areas and plains, and plains of river Indus and adjoining rivers in Sindh, Punjab and Khyber Pakhtunkhwa provinces [50].

It means with better inputs and management practices, average cane yields in Pakistan can be increased to 75 tons per hectare. In the present scenario, if the interest of millers prevails with somewhat more investment in cane yield maximization campaigns, 70 tons per hectare should be the national goal for Pakistan, and this would need extraordinary efforts.

7.1 Punjab

Punjab has very hot summer and very cold winter. Hot season starts by the month of April and continuous till August. The temperature is the highest (48–50°C) in extreme periods from May to July, with a mean maximum range of 37–42°C and a mean minimum range of 23–28°C for the corresponding period. During winter, the mean minimum range of 4–6°C is observed associated with occasional mild frost. Climate is more extreme in southwest regions in Multan and Khanpur zones. In monsoon season (Mid or late June), two-thirds of rainfall was received. The rainfall pattern of upper Punjab is higher and declines gradually toward central and southern Punjab [51]. Dera Ghazi Khan (D. G. Khan) is also situated in the dry region of the Northwest zone of Pakistan. Atmospheric humidity was low almost 33–40% during summer and 55–65% during winter due to low precipitation.

The cane area is concentrated in Faisalabad, Bahawalpur, Dera Ghazi, and Sargodha Divisions. The cane yields of Punjab have gradually increased to 64 t/h. However, within the province, average yields of 75 and 71 tons per hectare are obtained in Bahawalpur and D. G. Khan divisions, respectively. The average yield of Rajanpur District is reported to be 82 t/h, while the yield of R. Y. Khan is 77 t/h. The highest yield of 82.09 tons per hectare was obtained in Rajanpur district, followed by 77.75 tons per hectare recorded in R. Y. Khan district. The yield of districts located in Bahawalpur, D. G. Khan, and part of Faisalabad divisions are between 60 and 70 tons per hectare. In the rest of the districts, yields vary between 50 and 60 tons per hectare.

7.2 Sindh

Sindh is divided into two zones. The lower Sindh constitute Hyderabad and Mirpur Khas divisions, which are bordered by Arabian Sea coast. Compared with other regions, daily mean temperatures are relatively higher in winter and lower in summer, and the monthly mean shows lesser variability in maximum and minimum temperature. During May and June months, the mean maximum is higher (40–42°C), but the mean minimum (26–28°C) is favorable for growth and maturity phases.

In Badin and Thatta districts, temperature is relatively milder with cloudy weather during May–July. Coastal winds higher the humidity level as annual rainfall in Hyderabad (178 mm) and Badin (222 mm) districts is very low. Low relative humidity (46–66%) and mild mean minimum temperatures (8–16°C) free from frost during ripening are favorable for good sugar recoveries [16].

The upper Sindh, including Sukkur and Larkana divisions, is much away from the influence of coastal climate. The summer and winter months experience extreme weather with extremely low rainfall (88.2 mm). During summer, relative temperature is higher compared to lower Sindh, as such crops are subject to moisture stress. Stress conditions in Sukkur division are mitigated by better soil, crop, and water management practices. The yield of Sukkur division is the highest, around 75 t/ha, in Sindh followed by 67 t/ha in Benazir Abad. Larkana faces poor soils and more waterlogging; hence, yields of Larkana are much low. Climatically, the lower Sindh, including Bhanbore division, is close to the coast and should be more conducive to cane production than the central and upper Sindh areas. In fact, owing to higher cane and sugar yields, the cane fields of Thatta and Badin districts with part of Hyderabad and Mirpurkhas were declared the sugar land of Pakistan [16].

The highest yield of 82.7 tons per hectare was recorded in Ghotki district; Larkana and Nawabshah showed cane yields of around 69 tons per hectare. Eight districts showed a yield between 60 and 68 tons per hectare, while yields of the rest of the districts range between 40 and 60 tons per hectare. It may be noted with great concern that yields of districts Thatta and Badeen in the coastal area of lower Sindh have dropped down to 59.3 and 46.5 tons, respectively [16]. This is attributed to waterlogging with a large area under rice crop and fish farms.

7.3 Khyber Pakhtunkhwa

The province faces extreme climate during summer and winter months with mean maximum and minimum temperature ranges of 36–41°C and 4–5°C, respectively. Dera Ismail Khan (D. I. Khan) considered hot weather in summer with 271-mm rainfall annually, and Peshawar division considered cold weather in winter with 404-mm rainfall. Both the zones have a huge difference in rainfall patterns. This region has very low humidity and a higher RTD factor since there is no favorable effect on cane growth. The mean minimum temperature is low 4–5°C, and frosts are of common occurrence in the area [52].

Sugarcane is concentrated in Dera Ismail Khan, Peshawar, and Mardan divisions [52]. The cane yield of the province is hardly around 51 t/ha. However, Dera Ismail khan has attained a good position in cane yields (61.27 t/ha). D. I. Khan district is leading in cane yield with 62 t/ha. All of the remaining districts have yields lesser than 51 t/ha.

8. Cane area and production

Pakistan is a leading cane-growing country and ranks fifth in cane area and its production in the world. During the initial stage of the creation of Pakistan in 1947–1950, the cane was grown on an area of 201,900 hectares, and here, most of the cane was planted in Punjab (149,300 ha), a little in KPK (44,800 ha), and quite negligible in Sindh (7830 ha). There was one sugar factory in KPK and one very small factory in Punjab. Cane was mostly grown for cottage industry to produce *gur*, *shakkar*, and *khand* and only 2% of total cane production was utilized for white sugar production [6]. During the past seven decades, gradual expansion in sugar industry was associated with spontaneous expansion in the area under cane reaching to 1.039 million hectares, during 2020–2021 [29]. The decade-wise development in the area with cane yields during 1947–1950 to 2020–2021 period in cane-growing provinces of Pakistan is reproduced in **Table 4**.

Table 4 also indicates that in the initial periods of Pakistan existence (1950–1960), cane yields were around 26–33 tons per hectare. For the following 40 years (1960–1990), cane yields were oscillating just between 33 and 42 tons per hectare. Nevertheless, after 1990, cane yield started taking a boost with a gradual rise to 69.55 tons per hectare during 2020–2021. The Sindh province has even touched 65.56 tons per hectare during 2020–2021, while Punjab reached a level of 73.36 tons per hectare during 2020–2021. However, yields of KPK are still static, around 52.38 t/ha [52].

Medium and small farmers, those who own 2–10 acres of land, have a little flexibility to grow rice in addition to sugarcane. These farmers can purchase sugarcane inputs from private sources. Their primary concern is the timely availability of extension services from the public sector [1]. Increasing the share of sugarcane in these farmers' crop portfolios is the main goal of sugar mills.

Fast expansion in the sugar industry created competition in cane procurement that persuaded millers to initiate development activities in mills zones. A number of sugar mills invested in seed propagation of new cane varieties, fertilizers, and plant protection measures with technical guidance to growers [15, 19].

9. Sugarcane industry

Initially, cane was grown just to meet "*gur*," "*shakkar*," and "*khandsari*" demands of the local market and household needs [5]. With the creation of Pakistan in 1947, only two sugar mills (one in Sindh and one in Punjab) were operational. Resultantly, after every 10 years' period of span, we find a group of new sugar mills, scattered all around the cane-growing regions. At present, there are 89 sugar mills in operation, including 45 in Punjab, 38 in Sindh, and six in KPK.

In the initial period of industrial development, sugar mills installed crushing capacities were hardly around 1500–4000 tons' cane daily, but now daily cane-crushing capacities of sugar mills range from 12,000 to more or less 24,000 tons. The magnitude of the sugar industry expansion may be realized from the fact that during 1947–1950, the sugar mills had a crushing capacity of just 2800 tons of cane daily, while the daily cane-crushing capacity of the present-day sugar industry is around 590,000 tons, for 120-day crushing duration. Pakistan sugar industry has passed through different phases of development [19]. In its initial stage, almost all the cane was crushed in a local crusher for "*gur*," "*shakkar*," and "*khandsari*" production [6]. Only 2% of total cane production was utilized for white sugar manufacture in two small factories. Gradually, the sugar industry expanded to the extent that 95% of Sindh, 67% of Punjab, and 67% of KPK cane production were crushed in sugar mills [19]. In KPK, there is more trend of "*gur*" usage, and two sugar mills have been closed due to the nonavailability of cane for sugar mills [6].

10. Cane-crushing duration

In the earlier period of industrial development, while the crushing capacities of sugar factories were around 1500–4000 TCD, sugar mills operated for 180–240 days a year even at 3.5–5% recoveries in summer months. Mills used to start sometimes in October and continued crushing till June. Now, we find that sugar mills with usual crushing capacities of 8000–12,000 TCD still have several factories with 18,000–24,000 TCD. These days, cane is sufficient enough to complete crushing in a period of 120–135 days. The objective is to complete the crushing of cane during the period of its peak maturity. Nevertheless, Cane Act allows a crushing period of 1st November to 15th March for 160 days. While finalizing the crushing duration and especially the start of the crushing season, sugar recovery is of prime consideration.

11. Sugar recoveries

Sugar contents in cane play a leading role in regulating the economic viability of a sugar mill; in usual terms, it is indicated by "sugar recovery." It is the amount of sugar recovered per quintal (100 kg) of cane crushed in a sugar factory. It is affected by a number of factors in the environment, field, and factory [53, 54]. The matter has been discussed at length in the text; however, important ones are listed hereunder.

11.1 Factors affecting sugar recoveries

11.1.1 Cane varieties

Cane varieties have a dominant role in improving sugar mills recoveries. High-sugar-yielding varieties need to be grown, and preference should be given to early maturing varieties having high sugar contents.

11.1.2 Improving agronomic conditions

Improving agronomic conditions have a significant role in improving sugar mills recoveries. Following are some agronomic conditions that improve the sugar recovery of cane varieties.

- Good fertile salt-free land.
- Sweet water zone.
- Planting a time-longer growth period is preferred.
- Judicious use of balanced fertilizer, "K" application not to be ignored.
- Timely application of fertilizer, delayed application to be avoided and "N" contents in plant tissues should exhaust by maturity time.
- Judicious use of irrigation water, excess application to be avoided.
- Restrict irrigation before crop harvest.
- Measures to save the crop from lodging.

11.1.3 Natural calamities

Measures should be taken to save the crop from drought and frost.

11.1.4 Crop harvest strategies

Crop harvest schedule according to crop maturity and maturity period of cane varieties. Cane harvest according to the crushing capacity of sugar mills and milling operation to be regular and consistent. Cane stalks to be free of trash/extraneous matter; cane staling must be avoided.

11.1.5 Sugar factory milling and processing

All possible measures are to be taken to reduce sugar recovery losses during various milling and processing operations. The Pakistan sugar industry, during 65 years of development, has expanded from two small factories in 1947 to 89 sugar mills in

2016–2017. The first phase of 35–40 years is not much appreciable with respect to sugar recoveries attained in the factories. The sugar recoveries in Sindh province, due to its relatively favorable climate and somewhat better sugar varieties, have been observed at a little over 9%.

However, in Punjab, the major cane-producing area, the sugar recoveries have been oscillating between 8 and 8.5% up to the year 2000 [55]. The same is true for KPK province (**Table 8**).

The sugar industry has now a good number of quality varieties, including BL4, Thatta10, CP77-400, Nia 2004, NIA 2011, HS12, SPF234, CPF 237, CPF246, CPF 249, CPF 250, CPF 252, CPF 253, SG 676, Th 2109, and Th 326. These varieties have a sugar recovery range of 9.5–12.5%.

Resultantly, after the year 2010, the Pakistan sugar industry shows a progressive rise in its sugar recoveries, which have reached an average level of 9.61% during 2020–2021 (**Table 8**).

12. Conclusion

Sugarcane is the second most economically important crop of Pakistan after cotton. The data presented in this chapter showed that with the passage of time, the area, production, yield, and sugar recovery of sugarcane are increased greatly. However, in Pakistan, sugarcane is mostly used as multiple sources of food, fodder, and fertilizer as well as in many beverages as sweetener. A large number of people are involved with the sugar industry of Pakistan, thus creating human resource and work force for the country. The sugar industry of Pakistan earns huge revenue for the government. As for climatic conditions, Pakistan climate suits for its cultivation especially in Punjab, Sindh, and KPK provinces. In Punjab, Rajanpur and Rahim Yar Khan are the best suitable areas for sugarcane cultivation, while Sindh Ghotki and Thatta districts are the best suitable areas for sugarcane cultivation. In KPK, D. I. Khan and Murdan districts have the highest production. Sugar industry is gradually improving its status regarding its sugar recovery level. While checking the sugarcane varieties being grown and crushed in the factory, there appears a direct relationship between the cane varieties and the sugar recovery of the mills. The sugar mills having a large area under high sugar varieties would depict much better recoveries than the mills crushing a large percentage of low-quality cane. It is a matter of programming to harvest and crush cane according to its maturity, varietal combination matters much. The sugar industry of Pakistan needs a series of new varieties in the channel; for that matter, there should be a close liaison between the research institutes and the sugar industry for the betterment of sugarcane crop.

Author details

Shahid Afghan[1], Muhammad Ehsan Khan[1*], Waqas Raza Arshad[1], Karim Bukhsh Malik[2] and Amin Nikpay[3]

1 Sugarcane Research and Development Board, Ayub Agricultural Research Institute, Faisalabad, Punjab, Pakistan

2 Sugarcane Research Institute, Ayub Agricultural Research Institute, Faisalabad, Punjab, Pakistan

3 Department of Plant Protection, Sugarcane and By-Products Development Company, Salman Farsi Agro-Industry, Ahwaz, Iran

*Address all correspondence to: ehsankhanuaf@gmail.com

References

[1] Corcoran SG, Hashemi M, Sadeghpour A, Jahanzad E, Keshavarz RA, Liu X, et al. Understanding intercropping to improve agricultural resiliency and environmental sustainability. In: Sparks DL, editor. Advances in Agronomy. Vol. 162. Academic Press; 2020. pp. 199-256

[2] Qureshi S. Significance of sugar industry in National Economy. Econ. & Socl. Rev. 2004;**2**:17-21

[3] Annual Report of Pakistan Sugar Mills Association. Islamabad; 2020-21

[4] Chu CC, Kong L. Photo-respiration of sugarcane. Taiwan sugar Exp. Sta. Ann. Rept. 1971:1-14

[5] Qureshi MA, Afghan S. Sugarcane Cultivation in Pakistan. Pakistan Sugar Book. Karachi, Pakistan: Pakistan Society of Sugar Technologists; 2005

[6] Raza HA, Amir RM, Wudil AH, Usman S, Shoaib M, Ejaz R, et al. Economic analysis of jaggery (Gur) production in Tehsil Shakargar. Journal of Global Innovation in Agriculture and Social Sciences. 2018;**6**(2):69-73

[7] Dotaniya ML. Role of Bagasse and Press Mud in Phosphorus Dynamics. 1st ed. Germany: Lap Lambert Academic Publisher; 2014

[8] Prasannamedha G, Kumar PS, Mehala R, Sharumitha TJ, Surendhar D. Enhanced adsorptive removal of sulfamethoxazole from water using biochar derived from hydrothermal carbonization of sugarcane bagasse. Journal of Hazardous Materials. 2020;**407**:124825. DOI: 10.1016/j.jhazmat.2020.124825

[9] Fuhr. High or Low? The Sugar in Your Favorite Fruits. 2016. Available from: https://www.popsugar.com/fitness/Sugar-Content-Fruit-20134844

[10] Fahim MG. Study on yield and some agronomic traits of promising genotypes and lines of bread wheat through principal component analysis. Journal of Biological and Environmental Sciences. 2014;**2**:443-446

[11] Drewnowski A, France B. Liquid calories, sugar, and body weight. The American Journal of Clinical Nutrition. 2007;**2007**(85):651-661

[12] Chinnadurai C. Potential health benefits of sugarcane. In: Sugarcane Biotechnology: Challenges and Prospects 1-12. Springer International Publishing; 2017. pp. 1-12. DOI: 10.1007/978-3-319-58946-6_1

[13] Sahari J, Sapuan SM, Zainudin ES, Maleque MA. Physico-Chemical and Thermal Properties of Starch Derived from Sugar Palm Tree (Arenga pinnata). Asian Journal of Chemistry. 2014;**26**(4):955-959

[14] GOP. Pakistan Economic Survey (2019-20). Islamabad, Pakistan: Ministry of Food Agriculture and Livestock, Federal Bureau of Statistics, Government of Pakistan; 2019. pp. 27-34

[15] Raza HA, Amir RM, Saghir A, Tahir M. Sugarcane production and protection constraints faced by the growers of Punjab, Pakistan with special focus on the role of agricultural extension worker in related mitigation. Pakistan Journal of Agricultural Sciences. 2020;**57**(6):1681-1688

[16] Qureshi A, Sarwar PG, McCornick AS, Sharma BR. Challenges and prospects of sustainable

groundwater management in the Indus Basin. Pakistan. Water Resources Management. 2010;**24**(8):1551-1569

[17] Annual Report of Pakistan Metrological Department. Islamabad, Pakistan. 2019

[18] GOP. Pakistan Economic Survey (2020-21). Islamabad, Pakistan: Ministry of Food Agriculture and Livestock, Federal Bureau of Statistics, Government of Pakistan; 2020. pp. 27-34

[19] Iqbal MA, Iqbal A. Sugarcane production, economics and industry in Pakistan. American-Eurasian Journal of Agricultural & Environmental Sciences. 2014;**14**(12):1470-1477

[20] Malik KB. Agricultural and Industrial Aspects of Sugarcane Production. Lahore Pakistan: Al Madina Printers; 2018

[21] Nasir NM, Afghan S, Qureshi SA. Utilization of bio-compost produced from filter cake and stillage at Shakarganj sugar research institute, Jhang. Pakistan Sugar Journal. 1994;**8**:21-26

[22] Annual report of Shakarjang Sugar Mills, Jhang

[23] Usman M. Contribution of agriculture sector in the GDP growth rate of Pakistan. Journal of Global Economy. 2016;**4**(2):184-187

[24] Sugarcane Handbook. 2017. Shakarganj Sugar Mills Limited Jhang Pakistan

[25] Yasar A, Ali A, Tabinda AB, Tahir A. Waste to energy analysis of shakarganj sugar mills; biogas production from the spent wash for electricity generation. Renewable and Sustainable Energy Reviews. 2015;**43**:126-132

[26] Pakistan Agricultural Research Council. Sugarcane Crop in Pakistan, PARC, Islamabad. 2018. Available from: http://edu.par.com.pk/ wiki/sugarcane/

[27] Bhutta E, Ilyas M, Usman M. The need for transforming agriculture produce markets: Evidence from Punjab, Pakistan. Pakistan Journal of Agricultural Sciences. 2019;**56**(3):767-773

[28] Legendre BL, Martin FA. Ripening studies with Glyphosine in Louisiana sugarcane. Proceedings of American Society of Sugar Cane Technologists. 1977;**6**:62-64

[29] Annual Report of Pakistan Sugar Mills Association. Islamabad; 2016-17

[30] Luo J, Pan YB, Xu L, Grisham MP, Zhang H, Zhang H, et al. Rational regional distribution of sugarcane cultivars in China. Scientific Reports. 2015;**5**:15721

[31] Muhammad IT, Mohammad IJ, Iftekhar N, Naeem A, Abid M. A face for enhancing cane & sugar yield in Pakistan. Global Scientific Journal. 2019;**7**(3):670-686

[32] FAO. World Food and Agriculture—Statistical Yearbook 2017. Rome: FAO; 2017. DOI: 10.4060/cb1329en

[33] Marin FR, Edreira JIR, Andrade J, Grassini P. On-farm sugarcane yield and yield components as influenced by number of harvests. Field Crops Research. 2019;**240**(1):134-142

[34] Mendelsohn R. The impact of climate change on agriculture in Asia. Journal of Integrative Agriculture. 2014;**13**:660-665

[35] Kurukulasuriya P, Mendelsohn R, Hassan R, Benin J, Deressa T, Diop M, et al. Will African agriculture survive climate change? World Bank Economic Review. 2006;**20**:367-388

[36] Naoko U, Haruto S, Naohiro & Ryu O. Effects of the temperature

lowered in the daytime and night-time on sugar accumulation in sugarcane. Plant Production Science. 2009;**12**(4):420-427

[37] Das UK. Cane breeding in Coimbatore. Hawaiian Plant. Record. 1941;**45**:97-120

[38] Ali S, Liu Y, Ishaq M, Shah T, Ilyas A, Din IU. Climate change and its impact on the yield of major food crops: Evidence from Pakistan. Food. 2017;**6**(6):39

[39] FAO. World Food and Agriculture—Statistical Yearbook 2020. Rome: FAO; 2020. DOI: 10.4060/cb1329en

[40] Humbert RP. The Growing of Sugarcane. Amesterdam, London, NewYork: Elsevier Publishing Co.; 1968

[41] Hussain S, Khaliq A, Mehmood U, Qadir T, Saqib M, Iqbal MA, et al. Sugarcane production under changing climate: Effects of environmental vulnerabilities on sugarcane diseases, insects and weeds. Climate Change and Agriculture. 2018. DOI: 10.5772/intechopen.81131

[42] Biswass BC. Agroclimatology of the Sugarcane Crop (Technical Note N0: 193). Geneva, Switzerland: Secretariat of the world Meteorological Organization; 1988

[43] Afghan S, Jamil M. Climate change impact on sugar industry of Pakistan—An overview. Annual Convention. Pakistan Society of Sugar Technologists. 2013;**24**:7-14

[44] Kakade JR. Agricultural Climatology. New Delhi: Metropolitan Book. Co.; 1985

[45] Burr GO. The sugarcane plant. Annual Review of Plant Physiology. 1957;**8**:257-308

[46] Parthasarathy SV. Sugarcane in India. Mardras, India: K. C. P. Publishers; 1972

[47] Alexander AG. Sugarcane Physiology. Amsterdam: Elsevier; 1973. pp. 399-485

[48] Cardoza NP, Sentelhas PE. Climatic effect on sugarcane ripening under the influence of cultivars and crop age. Science in Agriculture. 1973;**70**(6):449-456

[49] Afghan S, Shahzad A, Comstock JC, Zhao D, Ali A. Registration of 'CPSG-3481' sugarcane. Journal of Plant Registrations. 2016;**10**:124-129

[50] Shakoor U, Saboor A, Ali I, Mohsin AQ. Impact of climate change on agriculture: Empirical evidence from arid region. Pakistan Journal of Agricultural Sciences. 2011;**48**(4):327-333

[51] Annual Program of Research Work. Sugarcane Research Institute. Faisalabad: Ayub Agricultural Research Institute; 2018

[52] Khan AQ, Muhammad I, Fahad I. An assessment of main problems faced by farming community in sugarcane production of district Peshawar. International Journal of Agricultural Extension and Rural Development. 2016;**3**(1):149-155

[53] Aslam C, Waqas M, Ahmad R, Khaliq A, Ahmad R. Improving the productivity and sugar recovery of cane by potash nutrition under different planting methods. Pakistan Journal of Agricultural Sciences. 2019;**55**(3):557-566

[54] Babu SC. Private sector extension with input supply and output aggregation: Case of sugarcane production system with EID-Parry in India. In: Zhou Y, Babu SC, editors. Knowledge Driven Development: Private Extension and Global Lessons. London, UK: Academic Press; 2015. pp. 73-90

[55] Afghan S, Hussnain Z. Clonal evolution program at Shakarganj sugar research institute, Jhang. Pakistan Sugar Journal. 2000;**25**(6):76-97

Chapter 5

The Sustainability of Sugarcane Ethanol in Brazil: Perspective and Challenges

Daniel Henrique Dario Capitani

Abstract

The chapter proposes to illustrate the challenges, concerns, and perspectives of ethanol production in Brazil. First, to give an overall of the sugarcane production and market conjecture, taking into account issues such as the public policies to promote biofuels improvement as well as those applied to energy markets and their connection (implications) with (into) sugar-ethanol market. Then, we propose the discussion of the challenges derived from sugarcane expansion from a sustainability perspective, as the environmental impacts, land use change and their impacts on crop productions and regional socioeconomics indicators, and the risk management strategies and tradeoffs between sugar-ethanol and second-generation ethanol-electricity cogeneration. Lastly, we bring a debate over the concerns and perspectives that are related to the development of this market, pointing out institutional risks that can affect strategies and competition in the production chain, such as policies to energy production, taxes changes, the increase in corn and sugarcane second generation ethanol production, and international trade agreements. Overall, there is an understanding that Brazilian ethanol production is following sustainable patterns. Currently, major challenges are related to the improvement of risk management strategies, as well as to create a more predictable scenario on the direction of public policies to the energy market.

Keywords: sugarcane, ethanol, sustainability, challenges, Brazil

1. Introduction

Since the colonial period in Brazil, sugarcane is cropped in the country and represents an important agricultural market, responding to the largest volume of sugar and ethanol in the international trade flows. Instead, ethanol production being started in the 1930s, the production reached a significant scale only in the 1970s, after the crude oil crisis and the positive shock on this commodity prices, as well as from a drop in the sugar prices. In 1975, the federal government set up the National Alcohol Program (ProAlcool), proposing to change fuels matrix in the country and reduce the dependence on gasoline prices. In the beginning, a blend of 20% of anhydrous ethanol was mixed with gasoline [1].

However, this new strategy required intense government participation on two sides. First, supporting institutions and subsidizing sugarcane mills, aiming to reach the minimal volume of ethanol production to attend the needs of fuels consumption (gasoline) in the domestic market. Thus, it was necessary to establish policies to improve sugarcane yield, with the adoption of new technologies in the sugarcane mills, fuels distribution and refining system and the management of the production mix between sugar and ethanol [1–3].

Second, government had to incentive new technologies in the vehicles industries for the adoption of technologies to turn the motors adapted to the blend of ethanol and gasoline and, further, to become them able to function using only hydrated ethanol [1–3]. This last issue requires an amount of investment (and subsides) in this industry for innovations that lead them to reach a significant share of domestic vehicles market with only-alcohol-fuelled cars. In this sense, beyond the subsidies given by Brazilian government to the automobile industry to turn feasible this type of vehicle, the innovation was possible by the engagement of several research centers and universities in the proposition of new technologies.

Regarding these government interventions, in the first half of the 1980s, hydrated ethanol and only-alcohol-fuelled car production exhibited a significant increase. The consequence of the increase in ethanol production was the large surplus of gasoline, which resulted in exports and incurred in investments to make changes in the crude oil refining structure. However, considering that ethanol is derived from an agricultural feedstock that has its own climate and biological uncertainties to the production flows, associated with the fact that sugarcane growers received low prices, the ethanol surplus exhibits large variations by each crop year. This fact, associated with the establishment of a large fleet of passenger cars fuelled only by ethanol, resulted in ruptures in this market and led to a return of consumers' preference for gasoline vehicles instead of only-alcohol [2, 3].

This movement kept intense until the 1990s and the Brazilian economic opening in 1991, that allow the imports of much advanced technological vehicles in comparison to only-alcohol-vehicles. The consequence of that was a strong reduction in the demand for passenger cars fuelled by hydrated ethanol and a persistent crisis in the sugarcane market, which became most dependent by sugar production and its prices fluctuation in international market [1–3]. This scenario had be maintained until 2003, when the flex-fuel injection technology was developed and started to be produced by automobile industries in Brazil, leading to a new expressive change in the country's sugarcane and biofuels market.

In order to illustrate the events that follow the expansion of sugarcane and ethanol production in Brazil from 2000s, this chapter proposes to explore how these changes impact the market increasing as well as the concerns and challenges that emerged in the past two decades. Therefore, the chapter has three sections. First, one explores the growth in Brazilian sugarcane production, addressing the initial investments boom and the sequent crisis in the sector. Second part of the chapter brings questions related to sustainability of sugarcane and ethanol production as this market experienced a substantial increase. Finally, third part investigates different concerns related to the institutional risks derived from public policies to energy market and new challenges that are emerging in sugarcane and ethanol markets.

2. The growth in sugarcane production

The increase in domestic hydrated ethanol consumption since 2003 had impact substantially the sugarcane cropping in Brazil. The country's average cropped land in the late 1990s and early 2000s was lesser than 5 million hectares. From 2003 to 2011, sugarcane area exhibits a progressive increase, reaching more than 10.5 million hectares. This growth has partially changed the distribution of sugarcane over the Brazilian territory. The state of Sao Paulo, traditionally producer, had kept the first position, responding for 55% of production. The growth had been also given by the participation of other states nearby Sao Paulo and in the Mid-West, such as Minas Gerais, Parana, Goias, Mato Grosso, and South Mato Grosso, increasing their share in the production area from 20% in 2000 to 35.8% in 2021. On the other hand, traditional producing areas in the North East had exhibited a drop from 22% in 2000 to 7.7% in 2021 [4, 5]. In **Figure 1**, we observe that sugarcane cropland and production expanded until 2010/11 crop year, oscillating from then, as the yield variations. **Table 1** exhibits the distribution of sugarcane production and harvested land by states in 2021/22 crop year.

This faster increase in sugarcane cropped land raise the average crop year production from 300 mi tons, in the late 1990s and early 2000s, to 630 mi tons from 2010 to 2020 (harvested land is 5% lower, on average). This current amount is twice the sugarcane production in India, the second largest producer of this commodity in the World [4, 6]. The production level consolidates Brazil as the largest sugar producer and exporter, and the second largest ethanol producer and major exporter in the World [6].

The importance of this agricultural market goes beyond its global importance in sugar and ethanol supply. Sugarcane bagasse and straw have been largely used in sugarcane mills in Brazil to cogenerate electricity since the middle of 2000s [7, 8]. Biomass was responsible for 8.8% of electrical energy generation in Brazil in 2021 and sugarcane bagasse/straw represent the largest amount of biomass source [9]. This is particularly important, considering the electricity in the country. Additionally, it has become a potential alternative for energy supply in the Brazilian South East, a region with 87 million people and responsible for 55% of the country's GDP, which in turn has been suffering with more severe drought over the past winters and overloading the hydroelectric system. As most of the sugarcane harvest in Brazil occurs in the winter and in this area, the biomass is an important alternative to the country's energy

Figure 1.
Sugarcane area and production in Brazil, 2005/06 to 2021/22 crop year [5].

State	Region	Production (million tons)	%	Area (millions ha)	%
Sao Paulo	Southeast	298,495	51.0%	4169	50.1%
Goias	MidWest	71,898	12.3%	963	11.6%
Minas Gerais	Southeast	63,948	10.9%	846	10.2%
South Mato Grosso	MidWest	44,180	7.5%	649	7.8%
Parana	South	31,962	5.5%	523	6.3%
Alagoas	Northeast	19,200	3.3%	308	3.7%
Mato Grosso	MidWest	15,292	2.6%	195	2.3%
Pernambuco	Northeast	12,648	2.2%	217	2.6%
Others		27,557	4.7%	447	5.4%
Total		585,179	100.0%	8317	100.0%

Source: [5].

Table 1.
Sugarcane production and harvested land by the Brazilian states in 2021/22 crop year.

planning. **Figure 2** presents the importance of sugarcane biomass for electricity production in Brazil, as other renewable energy sources.

Further, this market has an economic and regional importance, once there are more than 400 sugarcane mills in operation all over the country, many of them in small cities and being responsible for a significant share of local product. In general, the sector is the third in importance for the country's agribusiness, after the soybean and livestock market, and has been responsible for 1.2 million formal employments.

Finally, this market presented some changed in its economic structure. The international financial crisis in 2008/09 had impacts in this market, leading to the need of

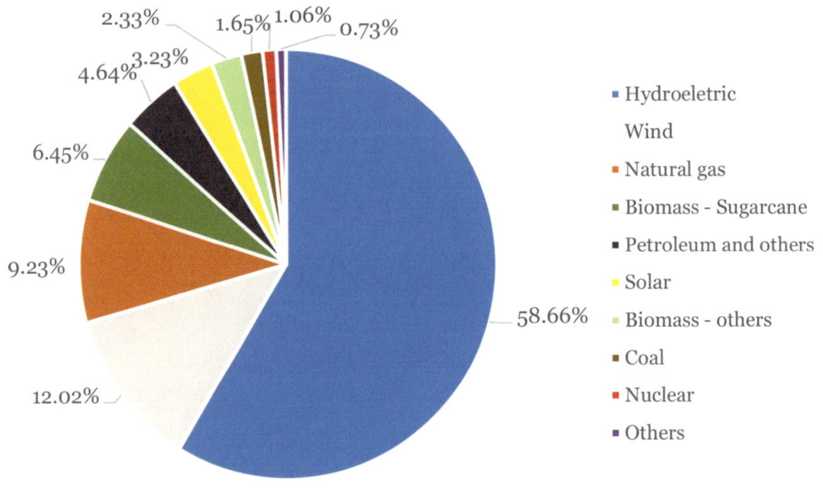

Figure 2.
Brazilian electricity generation by energy source, 2021. Source: [9].

financial restructuring in many groups of this industry, especially those that invested significant amount of capital to expand their industrial production capacity for sugar, ethanol and electricity cogeneration, dealing with a substantial increase over interest rate [10]. In addition, the consistent fall in the crude oil prices since 2008 and later, from 2011 to 2015, with the subsidies for fossil fuel consumption given by federal government through gasoline and diesel prices controlling had suppressed trade margins for sugarcane mills [11]. Finally, the agricultural expansion over nontraditional areas established a new learning curve and led sugarcane growers and mills to manage new agronomical practices in areas with different types of soil, weather, topography, and water resources [12]. All these facts had negative impacts over many of the established companies and, consequently, led to an increase in fusions and acquisitions, also attracting investments from external groups, most of them active in the petrochemical and infrastructure markets, with the participation of foreign companies.

Despite the evident changes in the macro and micro scenario of the Brazilian sugarcane market, there are still some issues that must be considered. Overall, it is possible to highlight three major issues that need more attention. First, how much the sustainability parameters have been attended by this production chain. In other words, it is necessary to find out if the sugarcane production has brought improvements in the environmental, social, and economic levels. A second question to be listed is the institutional relationship among the agricultural market agent's and the federal and state governments, evaluating the impacts of the public policies related to the environment, taxation, infrastructure, credit, electricity and fuels regulation and others, that may affect, in some way, the strategical decisions over the sugarcane market production, trade, and investments. Finally, the market reaching potential in the international trade and its challenges must be taken into account, as it can inhibit or stimulate the increasing of biofuels supply, especially considering the potential of Brazil to attend possible futures demand in the major markets, as European Unions and China, for instance.

Thus, some specific issues must be addressed. For example, the questions related to the land use change and indirect land use change, alternative biomass sources to produce ethanol, such as the own sugarcane bagasse or straw, or corn, considering the large increasing of this crop production in Brazil in more distant regions, whose production can be directed to local consumption. Therefore, the key is how to deal with such challenges, concerning with possible indirect impacts in the domestic production as well as the destination of ethanol or sugar for international trade.

3. The role of sustainability in the sugarcane production

After the ethanol supply expansion in the 2000s, some issues of concern related to sustainability have been highlighted, based on traditional methods applied to agricultural and industrial production. The main questions referred to the impacts resulting from the established process over the sugarcane production chain and how the solution for the unsuitable practices was being taken into account for sugarcane producers and mills, as well as for the sector institutions and government [13].

Such issues were addressed based on agricultural practices, as the use of water resources, the use of fertilizers, the sugarcane burn before the manual harvesting, and the new technologies for mechanical seeding and harvesting. For industrial efficiency, actions were taken to improve residues management and fermentation process.
For social and economic conditions, the reach for costs reduction regarding new

improvements in the agricultural and industrial process, the labor conditions, and the regional impacts of sugarcane production. And for general issues, the land use change resulted for production expansion, competition between food commodities production, and the possible impacts on biodiversity [13–23].

Regarding the use of fertilizers, sugarcane crop demands less fertilizer than other agricultural crop systems, especially in comparison to beet (destined for sugar in most of the developed economies) and corn (used for ethanol production in the USA, the largest producer in the World). In comparison to sugarcane crop in other countries, likewise Australia (160–200 kg N ha/year) and India (150–400 kg N ha/year), Brazilian production use significantly lower fertilizer (60–100 kg N ha/year) [22]. Furthermore, the recycling of nutrients by the use of two industries' residues led Brazilian production to a more sustainable level [17]. These residues are the vinasse (or stillage) and filter cake, derived from ethanol and sugar/ethanol production, respectively, managed as nutrient sources, optimizes according the agronomical and geographical conditions, and reducing the number of residues [15, 19]. At last, an important issue is related to how the vinasse has been treated in the sugarcane mills in comparison to the past, a large amount of these residues was not even treated or recycled, causing environmental problems, especially in the regional hydric resources. In addition, vinasse is useful in the irrigation of sugarcane land in the period of water shortage [13, 15, 19].

The use of water resources in the sugarcane crop had been pointed out since 1975, after the fast increase of the area covered by this commodity due to the ethanol production with the establishment of ProAlcool. Given the semi-perennial characteristics of this crop as well as the climatic regime of Sao Paulo state and nearby areas (responsible for 75% of production), with the concentration of rainfall in the initial phases of sugarcane growing, sugarcane cropping is favored and the irrigation is less necessary. However, the supplementary irrigation is needed in the Northeast and in some areas of the Mid-West, which in turn have been optimized by the use of fertirrigation using vinasse, allowing less use of traditional irrigation and lower use of mineral fertilizers. Further, water withdrawal was significantly reduced by sugarcane mills, from 5.6 $m^3 t^{-1}$ to 1.5 $m^3 t^{-1}$ between 1990 and 2008. Thus, these elements has shown that sugarcane is an agricultural activity more efficient than others in terms of the use of water resources, with the main cropped land using, in general, rainfall as the water resource in the field [13, 24, 25].

The sugarcane harvest was traditionally done through the burning process, aiming the costs reduction, once the burned biomass leads to a more accessible and feasible manual harvesting. However, this process results in an increase in carbon dioxide and other GHG emissions, which leads to a worsening of air quality in the surrounding cities [13, 15, 17]. However, since 2014, with the voluntary signature of the "Environmental Protocol" by sugarcane mills members of the Brazilian Sugarcane Industry Association (UNICA), a significant decrease was noted in the sugarcane burning in the State of Sao Paulo, and the mechanical harvesting exhibited a substantial increasing in Sao Paulo, as well as in the sugarcane expansion areas in the Mid-West. In 2008, less than 25% of Brazilian production was harvested mechanically, while in 2021 the number jump to almost 90% [5]. Take into consideration the major municipalities in the sugarcane in Sao Paulo State, it is remarkable the advancing in mechanical harvesting from 2007 to 2019 (**Figure 3**).

As a consequence of this changing, the global warming potential (GWP) of ethanol production in Brazil decreased by 46% and the black carbon (BC) emissions were seven times smaller when the non-burned harvesting reached 50% of the total area.

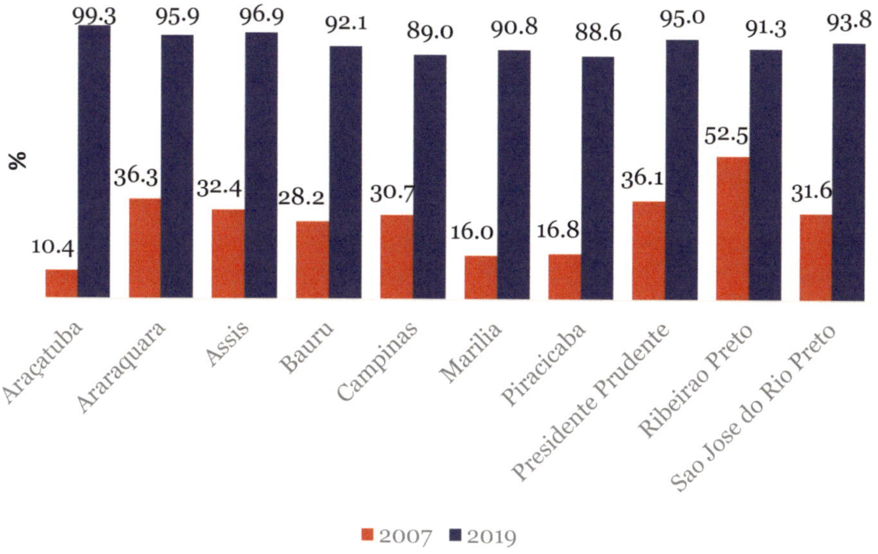

Figure 3.
Ratio of mechanical harvesting in sugarcane fields in Sao Paulo state, 2007 and 2019 [26].

Considering a scenario of 100% without sugarcane burning, the potential is the GWP decrease by 70% and BC will be 216 smaller [13].

Even in some areas with slope of 12% or higher, unapt for the mechanical harvesting, the protocol was adopted and the manual harvesting of green sugarcane has been taken into account. At last, two important issues may be pointed. First, mechanical harvesting leads to an increase in the amount of straw in the field, which can change the dynamics of soil erosion, soil nitrogen, soil water content and temperature, and soil carbon sequestration [17]. Unburned straw on soil can optimize the recycling of fertilizer nitrogen applied in the field as 105 kg ha/year [22]. In addition, an amount of this straw can be recovered and used in the electricity cogeneration or in the cellulosic ethanol production. Second, the change in mechanical harvesting has been demanding adjustments in agronomic management, such as the soil compaction and how to minimize the impacts on the yield and agronomic system [13, 22, 27].

Considering industrial efficiency, there is an effort to reduce the losses from fermentation with the improvement of distillation and dehydration technologies. Fermentation time reduced from 16 to 8 hours from 1975 to 2005 and the efficiency raised from 82 to 91% of the process [13]. Another important factor is related to investments on modern process of the mills' power section, especially recently, when electricity from biomass started to be traded in the spot electricity markets, as well as over renewable energy public [13]. The potential of power generation from sugarcane straw and bagasse is expressive, and can supply 15% of Brazilian energy demand [4]. This is most relevant considering that the mainly energy supply from sugarcane biomass occurs in the same period of the dry season in Southeast and Mid-West Brazil (area that includes 90% of domestic production), and is an alternative when the dam levels become to decrease and hydroelectrically energy prices rise.

The work conditions in the sugarcane activity have been taken into account since the rapid increasing of cropping after 1975. The concerns are related to the worker's health, labor journey, the payment by manual harvesting amount, the respect for labor legislation and the immigration of workers from the poorest areas (North and

Northeast) to work during the sugarcane harvesting. Indeed, these practices were not uncommon in the past, but from 2000s onwards a lot of progress was observed by improvements in the mills and their association's management, government laws and inspections, and the change in a mechanical harvesting system. Several studies point out an improvement in working conditions, with the creation of more qualified jobs in the field and industry (with higher payment), regional inequality reduction, a more restrict control in the labor journey as well as the establishment of properly places for rest and meals in the field [28–30]. In addition, the sugarcane sector exhibit a high ratio of workers formally registered in according to labor laws (95%), in comparison to 1981 (37.2%) and 2008 (81.4%), as well as with other traditional agricultural and services markets in Brazil [28]. However, there is still a concern related to the unemployment of workers in the manual harvesting, although some initiatives developed by mills association and public sector have been listed and applied in order to train and improve work skills of less qualified workers [31].

Several recent studies assessing the regional economic impacts of sugarcane activity in Brazil have been developed. Most of these studies used general equilibrium model or input/output models, simulating different scenarios considering the expected increase in sugarcane, ethanol, bioelectricity, and other factors and their direct and indirect effects on regional economic indicators, such as GDP and employment [20, 32, 33]. Overall, their suggestions have been pointing out that the positive effects on local economic activity tend to be greater in the expansion areas, where the growth in the harvesting areas can increase job offers and incur in the induced effects over other local trade activities. For example, considering a scenario of ethanol production increasing to 54 billion liters in 2030, there would be an increment of 2.6 billion USD and 53,000 new jobs nationwide [33]. In terms of GDP growth from 2000 to 2012, the municipalities with sugarcane processing increased their total real income by almost seven times, whereas the municipalities without sugarcane processing increased by less than six times [32]. In the traditional (and most economically developed areas) in Sao Paulo, the perspectives under new positive effects are more related with the dissemination of the second-generation ethanol (E2G). Additionally, in the Northeast, a development in the biomass electricity can significantly increase the employment and local GDP [20].

The discussion over socioeconomic and environmental impacts related to the sugarcane (ethanol) expansion in Brazil has started to consider the direct and indirect effects of land use change (LUC and ILUC) and the possible influences over other agricultural and livestock productions around the country. The area cropped with sugarcane had variated from 3 to 4% of the total current agricultural area, with half of all destined to ethanol production. Over the expansion period, from 2003 to 2011, the sugarcane area in Center South (responsible for 90% of the country's production) occurred mostly over pastures (close to 70%) and annual crops (25%). Forest areas, for instance, represented 0.6% of total area [13, 34, 35].

The most relevant empirical studies that proposed to examine the effect of biofuels (ethanol) on land use and food and fuel production in Brazil investigated the potential for transition in the livestock production practices and the impacts on the growth of sugarcane and ethanol production. Using partial equilibrium models and addressing data of agricultural, pastures and forest land distributions over the country, these studies simulated different scenarios of ethanol consumption and exports increasing, simultaneously with the dynamic of other agricultural and livestock markets. Overall, results suggest that in the scenarios with the greatest increase in ethanol production, sugarcane would expand from only 10 to 20% of the current area. Most of the area expansion would happen over existing pastures, which must assume a marginal increasing in intensification [34, 36, 37]. Thus, studies' findings suggest that even if ethanol production increase 100%, the direct impacts over other agricultural croplands would be little significant, with no risk to the country's food security.

The indirect effect of land use change (ILUC) and the hypothesis that the expansion of sugarcane indirectly causes deforestation in the Amazon forest is based on the fact that sugarcane intensification in the Center-South of Brazil contributes to an expansion of grain and livestock activities to North areas, surrounding the Amazon Forest. However, studies have shown that the dynamics between sugarcane, crops, and livestock are not clear and the presented hypothesis is a simplification and might be structured with strong arguments and data. The Brazilian Agroecological Zoning [38] points out to the existence of more than 60 million hectares currently available for these agronomical practices in Brazil, with no need for deforestation, e.g., corresponding for abandoned areas or degraded pastures suitable for sugarcane production. Further, both sugarcane and grains have shown significantly increasing on their yields in the past decades. Finally, the perspectives for an expansion in E2G ethanol may modify the previously listed hypothesis, once ethanol production could increase with no additional of the cropped area [13]. It is also important to point out that most of the sugarcane area is spread in the traditional areas, where grains production area not disseminated. Even in the Midwest, traditional grain producer, sugarcane area is less than 5% of soybean cropland [5].

The increasing production of biofuels has also promoted a debate on the effects of biofuels on food prices. A special concern with development countries is taking into account, once an increase in food commodities prices might affect the low-income population and evidence the problem of food security [39]. Thus, the relationships between food, energy, biofuels, and commodities prices have become a pertinent topic for discussion and improve the analysis related to the sustainability of biofuels production [40].

Few studies focused their analysis on understanding the impacts of ethanol (and sugarcane) expansion on food prices. Prior studies proposed to investigate their analysis to assess price and volatility linkages between ethanol, gasoline, oil and sugar prices [40] in the U.S. and international markets, suggesting that the volatilities among international crude oil and sugar prices have significant impacts in the dynamics of ethanol prices, even in Brazil [39, 41–46]. However, there are no evidences of opposite causality effects, e.g., that ethanol has any influence over energy prices. In the international market, this effect was tested by U.S. corn ethanol, but only in few periods the evidences point out to causality in the ethanol prices to the U.S. corn prices, although any other flows were identified, as, for example, impacts in the U.S. gasoline or other grain prices [43–47]. For Brazil, previous studies includes sugar prices together with energy prices and found that sugar can influence ethanol prices, as well as gasoline and oil prices, but no effects were identified in the other side [39–42, 46].

In addition, the Brazilian federal government policies in the fuels market (gasoline) represent a limitation on the domestic ethanol price fluctuations. More recently, new studies included more variables in the time series models, as other food commodities prices (soybean, corn, rice, cattle, wheat, and cassava), to measure the indirect effects, as well as exchange rate, which can affect the determinants of commodities exports or imports. The results suggest that none of these prices are significantly affected by ethanol or sugarcane prices. However, crude oil and exchange rates can influence commodities prices, once they affect the cost of production and revenue in these markets [47–50]. Thus, there are no evidences that ethanol production in Brazil can affect the dynamics of domestic food commodities prices. In association with the findings from the studies that assessed the LUC and ILUC derive from sugarcane production in Brazil, these results show that food production systems in Brazil were affected by the boom in ethanol consumption in the country, as well as there are no evidences that this status can be modified in the short run.

4. Risk management, public policies, and institutional impacts over Brazilian sugarcane and ethanol production

Another important issue related to the production of ethanol in Brazil is the particular structure of sugarcane sector, whereas sugarcane is the major feedstock for both sugar or ethanol production as well as sugarcane straw and bagasse from the sugar/ethanol production are inputs for electricity production by many of the sugarcane mills. Thus, from one side, the decision is strongly related to the dynamics of sugar prices in the international market [51, 52], simultaneously by the dynamics in the fuels market in Brazil and World. On another side, although it does not influence the sugarcane (feedstock) crushing and the decision over the mill's production mix, the energy cogeneration approaches the interrelationship between this sector with major energy market, once it affects the costs and revenues of sugarcane mills [53].

From this productive conjecture, it is evident that strategies for ethanol production in Brazil are influenced from other markets, such as sugar, fuels, and energy. Therefore, uncertainties emerge in the both production and trade decisions, demanding governance, production, and trade optimization in the production chain. In addition, these uncertainties raise by inherent agricultural and market risks, as well from institutional risks derived by government policies related to the agriculture, fuels, and energy markets [54].

The challenges over the production mix decision include the current information and forecast for sugar and ethanol prices, the global and local demand by food industries and fuel consumers, the agricultural costs and yield of sugarcane, industrial costs of sugar and ethanol process, operational costs, and scale efficiency [55]. Thus, strategies are conducted according to ethanol and sugar stocks and prices, gasoline prices, electricity prices, cost of production for sugarcane, sugar and ethanol, conjecture of the international market (especially for sugar), and the mills cash flows needs. Overall, the decisions are taken at the beginning of the sugarcane harvesting and can be different yearly [56].

Uncertainties related to sugar and ethanol prices in Brazil require complex financial and productive strategies for sugarcane mills. Consequently, they need to develop optimized strategies for both industrial production as well as for trading these outputs [57]. Portfolio optimization demands a broad knowledge of stocks, production and trade. Simultaneously, sugarcane mills need to manage their financial strategies in the short run. For that, this market has to deal with production and price risk for feedstock (sugarcane) and their outputs (sugar and ethanol) [53].

This association between agriculture and price risks increases the challenges for the decision makers in this market. Sugarcane is an annual crop, and full harvest period in Brazil is from April to November. Thus, in a scenario of disequilibrium between sugar and ethanol production, concomitantly with the volatility of their international prices, can increase risks for the mills, whereas profit margins can move in different directions over the harvest period, worsening financial results for the industry [58]. For example, **Figure 4** shows the difference between sugar and ethanol monthly average profit margins over 2004–2014 collected by a sample of more than 120 sugarcane mills around Sao Paulo, Parana, Minas Gerais, Goias and South Mato Grosso. In this period, sugar prices exhibited a significant increasing worldwide, while domestic ethanol prices were dependent by gasoline price variation. As consequence, ethanol price risks seem to be higher than sugar price risks, especially in periods of government interference in gasoline prices [11, 54].

On the other side, institutional risks can derive from changes in policies related to sugarcane and energy markets that can be associated with tax changes and interventions in prices and trade. For the energy co-generation, investments were stimulated by Brazilian government in the National Planning for Agroenergy (PNA), National Planning for Energy 2030 (PNE), and Decennial Plan to Energy Expansion (PDE) [59]. However, the global economic crisis in 2008 increases interest rates, reducing investments. Additionally, considering that Brazilian government bioenergy auction did not differentiated electricity sources, a scenario of increasing in sugarcane mills costs [60] reduces the competitiveness of sugarcane bagasse and straw energy co-generation in comparison to other bioenergy sources, such as wind power, that received most part of investments along 2010 decade [61].

The global economic crisis in 2008 also affect the crude oil prices, which sharply decreased fossil fuels costs, such as gasoline and diesel, decreasing the competition for biofuels, such as ethanol. In Brazil, this event was associated with an increase in consumer price indexes, which was the trigger to changes in federal government policies for energy markets, controlling gasoline and electricity prices, aiming the reduction of the inflation rate. The price controlling policy is common in Brazil, since the country has a history of periods with high inflation rate. Even in the periods of regular price indexes, fuels prices are not entirely free in the country, once all crude oil derivatives in Brazil is set by the public company Petrobras [11]. When this practiced policy is highly interventionist, as followed between 2011 and 2015, there are significant divergences of domestic fuel prices from the international price. As consequence, ethanol prices, which are mostly derived from agricultural and industrial sugarcane

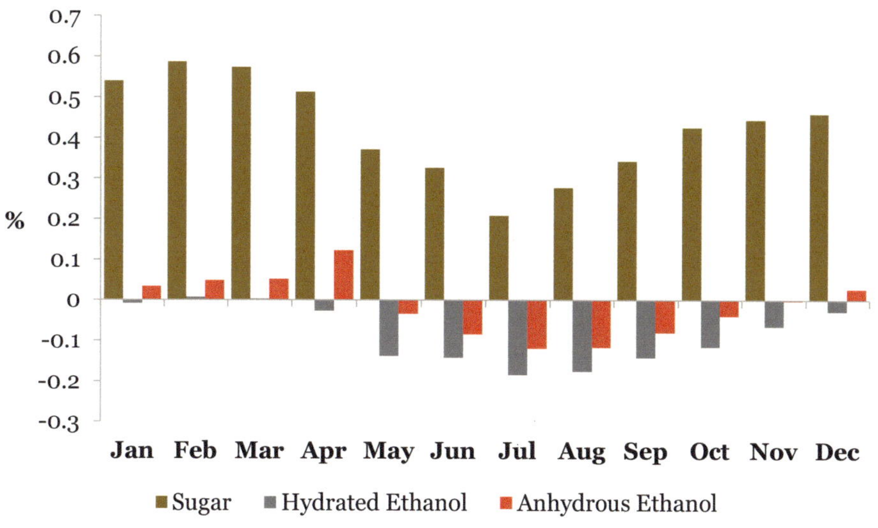

Figure 4.
Monthly average profit margins for sugar, hydrated and anhydrous ethanol in Brazil, 2004–2014 [58].

mills' cost of production, exhibit a barrier for their free fluctuation when the costs and prices go up. This practice also results in a lack of new investments in the oil-related (and biofuel) sectors and fuel supply shortages, and it affects the profitability of Petrobras and the government debts [62]. **Figure 5** points out ethanol and gasoline prices for consumers in Sao Paulo state. The shadow area represents the period of intervention. As result, ethanol prices positive fluctuations are restricted and, consequently, its volatility is lower. Over the period before (after) intervention, volatility of gasoline and hydrated ethanol was 0.37 and 0.45 (0.68 and 0.64), and in the period of intervention, volatility was 0.19 and 0.26, respectively. Still, it is necessary to consider that prices seem to be more variable along in the intervention period, once the data are in real terms, e.g., they were evaluated by the consumer price index.

The recent period of high intervention in the fossil fuels market in Brazil led to major impacts for the Brazilian sugarcane industry, such as the increase in their cost of production, a decrease in the sector profit margins and difficulties to achieve financial credit. In this sense, the adoption of free prices in gasoline market in Brazil can positively affect the dynamics of biofuel market in the country [10, 11]. The current policy of Petrobras, adopted from 2016 to this current period is much more directed for the free market, although the fact of Petrobras act as a monopoly in the fuels refining in the country is an element that shed light that this market is not entirely free.

However, this practice is not guarantee in the short run, considering the impacts in the crude oil prices, as consequence of the Covid-19 pandemic effects in the global production chain, as well as the impacts of the conflict between the Russian Federation and Ukraine. Brazilian federal government has been requiring a change in the Petrobras prices policies for diesel and gasoline, although the company and stakeholders are still supporting the current free prices policies.

Despite the conjectural issues affecting the market in the past decades, especially from the massive intervention over energy prices, increasing uncertainties, and decreasing investments, Brazilian sugarcane market has also other challenges to overcome, related to structural issues, as the low increase in crop yield and the fast dissemination of mechanical harvest [12]. The first concern results in low levels of

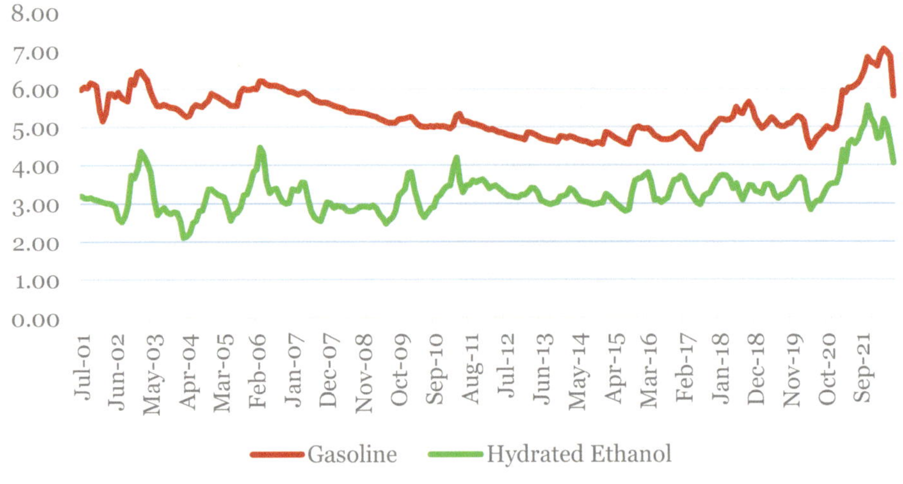

Figure 5.
Monthly gasoline and ethanol prices in Sao Paulo State, Brazil, 2001–2022, R$/liters (prices in real terms, July/2022) [63].

research and development in this market in comparison to other crops, once the significantly smaller production volume. The second issue is a consequence from the fast adoption of mechanical harvest in the country fields, requiring an improvement in the learning curve of sugarcane producers to avoid soil compression. This can be noted by the decrease in sugarcane yield, from an average of 81.3 ton/ha in the late 2000s, to 73.3 in the late 2010s [5].

In addition, institutional risks may come from the relationship of Brazilian ethanol in the international market. Locally, from the recent corn ethanol production increasing, that could affect the classification of Brazilian sugarcane ethanol as advanced fuel. In 2021/22, 13% of Brazilian ethanol was derived from corn, and the estimative pointed to a volume of 20% in 2030 [4].

This increase is resulted of the expansion of winter corn in Brazil in the past two decades, especially in the Mid-West, an area far from the major sugarcane cultivated areas and from the major ethanol consumption market. This region has a large corn production, and the use of this product as feedstock for ethanol production is positive for local producers and industries, once the sugarcane ethanol is costly locally, whereas corn stocks tends to increase when prices go down. Therefore, the supply of ethanol production locally is less costly and allows a better risk management for corn producers, traders and industry in Mid-West, reducing economic impacts in this production chain and benefiting local ethanol consumers [64, 65]. Therefore, in a context that Brazil needs to gauge new markets, there is a need to guarantee that corn ethanol production is not in competition with sugarcane ethanol and is not a threat to the country position as exporter of an advanced product.

Externally, uncertainties come from the conduction of major countries' policies and mandates for the use of renewable fuels, especially the EU and China. Although U.S. and Brazil have been dominants in the ethanol international market, both countries are mostly trading with each other, depending from several variables that affects their stocks and production, as the prices of feedstocks, crude oil and sugar, the domestic demand and harvesting issues [66]. The expansion of international trade is dependent from the reduction of trade tariffs and the adoption of the intended mandate of biofuels by these countries. In China, the ethanol mandate for 11 provinces has been expanding the use of this biofuel, but the effects are still lower on the Brazilian exports [67]. However, this level can be increased by the adoption of similar policies in many of Asian markets, such as Philippines, India, and Vietnam. For European Union, the current policies did not incur to significant volumes of exports for Brazilian ethanol, but the New Green Deal and REDII set ambitious targets for renewable fuels [68] and new agreements between Brazil and EU could be positively for sugarcane production chain.

At last, another current challenge for this sector is the advancement in the second generation ethanol (E2G), which can affect the strategies in the electricity cogeneration, as well in the sugar ethanol production from the original sugarcane feedstock. In Brazil, the focus is in the subvention, support to research, and development of technological routes to promote investments and guarantee their feasibility. Thus, the focus is to promote economic sustainability for suppliers, involving all the production chain, which may be more effective before convincing potential consumers [64].

Based on this, initiatives promoted and supported by the government and private sector have been conducted between industries, universities and research centers, such as the BNDES-Finep Support for Industrial Technologic Innovation for Sugarcane Sector (PAISS) and Fapesp Bioenergy Program Research (BIOEN) [63, 69]. Currently, only few sugarcane mills in Brazil have adopted the technologies to produce E2G. Some of them reduced the volume after the beginning, due to the high costs or technical inefficiency, which require more time for the project maturity and the establishment of policies to support the investments, to promote a regular production, and to stimulate the consumption [69].

Therefore, the development of E2G in Brazil is dependent on the role of new technologies, energy policies, and investment availability [69, 70]. The efforts to maintain a suitable governance is conducted by the RenovaBio, a national policy for biofuels that established goals to mitigate carbon emissions in the fuels market and to promote the use of biomass in the country energy matrix. The major commitments of RenovaBio are the achievement of Paris Agreement goals; contributing to reducing GHG emissions and improving energy efficiency; improve the management of biofuels production, expansion and use; offer frameworks to manage the supply of several fuels in the domestic market; and improve the sustainability in the energy market. Moreover, RenovaBio introduces an innovative approach to creating and developing a market for carbon credit (CBIOs) [71–73].

Besides the concerns over the environment and energy security, RenovaBio brings a product that can attract new investors to this market, once it can be negotiated in the stock exchanges, as a forward contract. CBIO is calculated by the difference between fossil fuel CO_2 emission and its biofuel substitute. The higher this difference, the more CBIO can be traded. Therefore, as E2G generates a smaller carbon footprint compared to first-generation ethanol, it can stimulate E2G viability. Thus, RenovaBio can potentially reduce ethanol prices and promote the expansion of advanced biofuel technology development and production in Brazil [72].

5. Conclusions

This chapter proposes to address the challenges, concerns, and perspectives of bioethanol production in Brazil. From a market conjecture perspective, sugarcane cropping had a significantly increasing from 2003 to 2010 as a result from the implementing of flex-fuel vehicles in Brazil by 2002, which increased the demand for hydrated ethanol. The commodities markets boom also stimulated this movement over 2000 decades, where Brazil raise its sugar and ethanol exports, consolidating its position as a great player in both markets.

From this event, an extensive debate was carried out about the sustainability of ethanol production in Brazil, mainly based on the old practices adopted in the field, such as sugarcane burning, soil and water pollution, the working conditions and the impacts on food commodities prices. However, several studies conducted so far have been demonstrating that the sugarcane expansion in Brazil is quite well suited for the environmental, social and economic issues, following the precepts of sustainable development.

Overall, studies point out that environmental issues are under control. Residues have been treated in the mills, there is a short use of water resources in the cropped fields, the mechanical harvesting advanced in most part of the production area, reducing burning and promoting a tillage cropping system with the straw, reducing the nutrients losses and retaining carbon soil. Besides, from social side, there was substantial improvement in the working conditions. From the economics side, investments impact positively the regional GDPs and employment rate, where the LUC and ILUC have no significant impact in the food production or environment, as well as ethanol has no impact on other important commodities produced in the country, even if considering geographical crops distributions or food prices.

However, after the remarkable increase and the improvement of sustainability aspects, there are challenges that Brazilian sugarcane market has to manage in the coming years. Economic and financial skills need to be improved to deal with sugar and ethanol production mix and trade, optimizing strategies over a crop year. Still, the decisions are subsidized by political and institutional issues, such as the intervention in fuel markets and the public auctions for renewable energy sources, such as sugarcane biomass. Nevertheless, other institutional issues can affect the market strategies, such as biofuel mandates in Brazil and other countries, and the linkages in their trade policies, and the CBIOs trade by Brazilian producers to mitigate GHG emissions. Lastly, the advancing in corn ethanol and E2G production can potentially change current production and trade in this market and bring new elements to deal with.

Author details

Daniel Henrique Dario Capitani
School of Applied Sciences, University of Campinas, Limeira, SP, Brazil

*Address all correspondence to: danieldc@unicamp.br

References

[1] Rosillo-Calle F, Cortez LAB. Towards proalcool II—A review of the Brazilian bioethanol programme. Biomass and Bioenergy. 1998;**14**(2):115-124. DOI: 10.1016/S0961-9534(97)10020-4

[2] Soccol CR, Vandenberghe LPS, Costa B, Woiciechowski AL, Carvalho JC, Medeiros ABP, et al. Brazilian biofuel program: An overview. Journal of Scientific & Industrial Research. 2004;**64**:897-904

[3] Stolf R, Oliveira APR. The success of the Brazilian alcohol program (Proálcool)—A decade-by-decade brief history of ethanol in Brazil. Engenharia Agrícola. 2020;**40**(2):243-248. DOI: 10.1590/1809-4430-eng.agric.v40n2p243-248/2020

[4] Unica—Brazilian Sugarcane Industry Association. Observatório da Cana. 2022. Available from: http://www.unicadata.com.br. [Accessed: June 27, 2022]

[5] Conab—National Supply Company. Safras: Series históricas da cana-de-açúcar. 2022. Avalilabe fom: http://www.conab.gov.br. [Accessed: June 27, 2022]

[6] USDA. Foreign Agricultural Service. Data and analysis. 2021. Available from: http://www.fas.usda.gov. [Accessed: May 17, 2022].

[7] Dias MOS, Modesto M, Ensinas AV, Nebra SA, Maciel Filho R, Rossell CEV. Improving bioethanol production from sugarcane: Evaluation of distillation, thermal integration and cogeneration systems. Energy. 2011;**36**(6):3691-3703. DOI: 10.1016/j.energy.2010.09.024

[8] Dias MOS, Junqueira TL, Cavalett O, Cunha MP, Jesus CDF, Mantelatto PE, et al. Cogeneration in integrated first and second generation ethanol from sugarcane. Chemical Engineering Research and Design. 2013;**91**(8):1411-1417. DOI: 10.1016/j.cherd.2013.05.009

[9] Brazil. Brazilian Electricity Regulatory Agency. 2022. Available from: http://https://www.gov.br/aneel/. [Accessed: July 04, 2022]

[10] Santos GR, Garcia EA, Shikida PFA. A crise na produção do etanol e as interfaces com as políticas públicas. Radar. 2015;**39**:27-28

[11] Popova ND. Fuel and biofuel sectors in Brazil - comparison with developed economies and analysis of hypothetical free fuel pricing policy [thesis]. Piracicaba, Brazil: University of Sao Paulo; 2017. DOI: 10.11606/T.11.2017.tde-29092017-160029

[12] Nyko D, Valente MS, Milanez AY, Tanaka AKR, Rodrigues AVP. A evolução das tecnologias agrícolas do setor sucroenergético: Estagnação passageira ou crise estrutural? BNDES Setorial. 2013;**37**:399-442

[13] Walter A, Galdos MV, Scarpare FV, Leal MRLV, Seabra JEA, Cunha MP, et al. Brazilian sugarcane ethanol: Developments so far and challenges for the future. WIREs Energy and Environment. 2014;**3**:70-92. DOI: 10.1002/wene.87

[14] Goldemberg J, Coelho ST, Guardabassi P. The sustainability of ethanol production from sugarcane. Energy Policy. 2008;**36**(6):2086-2097. DOI: 10.1016/j.enpol.2008.02.028

[15] Galdos MV, Cerri CC, Lal R, Bernoux M, Feigl B, Cerri CEP. Net greenhouse gas fluxes in Brazilian

ethanol production systems. Global Change Biology Bioenergy. 2010;**2**:37-44. DOI: 10.3390/su4040574

[16] Cavalett O, Cunha MP, Junqueira TL, Dias MOS, Jesus CDF, Mantelatto PE, et al. Environmental and economic assessment of bioethanol, sugar and bioelectricity production from sugarcane. Chemical Engineering Transactions. 2011;**25**:1007-1012. DOI: 10.3303/CET11255168

[17] Galdos M, Cavalett O, Seabra JEA, Nogueira LAH, Bonomi A. Trends in global warming and human health impacts related to Brazilian sugarcane ethanol production considering black carbon emissions. Applied Energy. 2013;**104**:576-582. DOI: 10.1016/j.apenergy.2012.11.002

[18] Duarte CG, Gaudreau K, Gibson RB, Malheiros TF. Sustainability assessment of sugarcane-ethanol production in Brazil: A case study of a sugarcane mill in São Paulo state. Ecological Indicators. 2013;**30**:119-129. DOI: 10.1016/j.ecolind.2013.02.011

[19] Leal MRLV, Galdos MV, Scarpare FV, Seabra JEA, Walter A, Oliveira COF. Sugarcane straw availability, quality, recovery and energy use: A literature review. Biomass and Bioenergy. 2013;**53**:11-19. DOI: 10.1016/j.biombioe.2013.03.007

[20] Martínez SH, van Eijck J, Cunha MP, Guilhoto JJM, Walter A, Faaij A. Analysis of socio-economic impacts of sustainable sugarcane–ethanol production by means of inter-regional Input–Output analysis: Demonstrated for Northeast Brazil. Renewable and Sustainable Energy Reviews. 2013;**28**:290-316. DOI: 10.1016/j.rser.2013.07.050

[21] Filoso S, Carmo JB, Mardegan SF, Machado SRL, Figueiredo TG, Martinelli LA. Reassessing the environmental impacts of sugarcane ethanol production in Brazil to help meet sustainability goals. Renewable and Sustainable Energy Reviews. 2015;**52**:1847-1856. DOI: 10.1016/j.rser.2015.08.012

[22] Bordonal RDO, Carvalho JLN, Lal R, Figueiredo EB, Oliveira BG, La Scala N Jr. Sustainability of sugarcane production in Brazil. A review. Agronomy for Sustainable Development. 2018;**38**(13):12-23. DOI: 10.1007/s13593-018-0490-x

[23] Picoli MCA, Machado PG. Land use change: The barrier for sugarcane sustainability. Biofuels, Bioproducts and Biorefineries. 2021;**15**:1591-1603. DOI: 10.1002/bbb.2270

[24] Scarpare FV, Hernandes TAD, Ruiz-Corrêa ST, Picoli MCA, Scanlon BR, Chagas MF, et al. Sugarcane land use and water resources assessment in the expansion area in Brazil. Journal of Cleaner Production. 2016;**133**:1318-1327. DOI: 10.1016/j.jclepro.2016.06.074

[25] Guarenghi MM, Walter A. Assessing potential impacts of sugarcane production on water resources: A case study in Brazil. Biofuels, Bioproducts & Biorefining. 2016;**10**:699-709. DOI: 10.1002/bbb

[26] Fredo CE, Caser DV, Campagnuci BCG. Colheita mecanizada da cana-de-açúcar atnge 95.3% das áreas produtivas do estado de São Paulo na safra agrícola 2018/19. Análise e Indicadores do Agronegócio. 2020;**15**(7):1-9

[27] Capaz RS, Carvalho VSB, Nogueira LAH. Impact of mechanization and previous burning reduction on GHG emissions of sugarcane harvesting

operations in Brazil. Applied Energy. 2013;**102**:220-228. DOI: 10.1016/j.apenergy.2012.09.049

[28] Moraes MAFD. O mercado de trabalho da agroindústria canavieira: Desafios e oportunidades. Economia Aplicada. 2007;**11**(4):605-619. DOI: 10.1590/S1413-80502007000400008

[29] Moraes MAFD, Oliveira FCR, Diaz-Chavez RA. Socio-economic impacts of Brazilian sugarcane industry. Environmental Development. 2015;**16**:31-43. DOI: 10.1016/j.envdev.2015.06.010

[30] Balsadi OV. Mercado assalariado na cultura da cana-de-açúcar no período de 1992-2006. Revista de Economia Agrícola. 2010;**57**:91-110

[31] Hall J, Matos S, Severino L, Beltrão N. Brazilian biofuels and social exclusion: Established and concentrated ethanol versus emerging and dispersed biofuels. Journal of Cleaner Production. 2009;**17**(11):577-585. DOI: 10.1016/j.jclepro.2009.01.003

[32] Caldarelli CE, Moraes MAFDD, Paschoalino PAT. Sugarcane ethanol industry effects on the GDP per capita in the Center-South region of Brazil. Revista De Economia e Agronegócio. 2017;**15**(2):183-200. DOI: 10.25070/rea.v15i2.481

[33] Brinkman MLJ, Cunha MP, Heijnen S, Wickle B, Guilhoto JJM, Walter A, et al. Interregional assessment of socio-economic effects of sugarcane ethanol. Renewable and Sustainable Energy Reviews. 2018;**2018**(88):347-362. DOI: 10.1016/j.rser.2018.02.014

[34] Nassar AM, Harfuch L, Bachion LC, Moreira MR. Biofuels and land-use changes: Searching for the top model. Interface Focus. 2011;**1**:224-232. DOI: 10.1098/rsfs.2010.0043

[35] Adami M, Rudorff BFT, Freitas RM, Aguiar DA, Sugawara LM, Mello MP. Remote sensing time series to evaluate direct land use change of recent expanded sugarcane crop in Brazil. Sustainability. 2012;**4**:574-585. DOI: 10.3390/su4040574

[36] Khanna M, Crago CL, Black M. Can biofuels be a solution to climate change? The implications of land use change-related emissions for policy. Interface Focus. 2011;**1**:233-247. DOI: 10.1098/rsfs.2010.0016

[37] Nuñez HM, Önal H, Khanna M. Land use and economic effects of alternative biofuel policies in Brazil and United States. Agricultural Economics. 2013;**44**(1):487-499. DOI: 10.1111/agec.12032

[38] Manzatto CV, Assad ED, Bacca JFM, Zaroni MJ, Pereira SEM, editors. Zoneamento agroecológico da cana-de-açúcar. Rio de Janeiro: Embrapa Solos; 2009. p. 55

[39] Sexton S, Rajagopal D, Zilberman D, Hochman G. Food versus fuel: How biofuels make food more costly and gasoline cheaper? Agricultural and Economic Resource Update. 2008;**12**(1):1-6

[40] Serra T, Zilberman D. Biofuel-related price transmission literature: A review. Energy Economics. 2013;**37**(1):141-151. DOI: 10.1016/j.eneco.2013.02.014

[41] Rapsomanikis G, Hallam D. Threshold cointegration in the sugar-ethanol-oil price system in Brazil: Evidence from nonlinear vector error correction models. FAO Commodity and Trade Policy Research Working Paper. 2006;**22**:1-14

[42] Balcombe K, Rapsomanikis G. Bayesian estimation and selection of nonlinear vector error correction models: The case of the sugar-ethanol-oil nexus in Brazil. American Journal of Agricultural Economics. 2008;**90**(3): 658-668. DOI: 10.1111/j.1467-8276.2008.01136.x

[43] Zhang Z, Lohr L, Escalante C, Wetzstein M. Food versus fuel: What do prices tell us? Energy Policy. 2010;**38**(1):445-451. DOI: 10.1016/j.enpol.2009.09.034

[44] Timilsina G, Mevel S, Shrestha A. Oil price, biofuels and food supply. Energy Policy. 2011;**39**(1):8098-8105. DOI: 10.1016/j.enpol.2011.10.004

[45] Serra T, Zilberman D, Gil JM, Goodwin BK. Price volatility in ethanol markets. European Review of Agricultural Economics. 2011;**38**(1):259-280. DOI: 10.1093/erae/jbq046

[46] Zilberman D, Hochman G, Rajagopal D, Sexton S, Timilsina G. The impact of biofuels on commodity food prices: Assessment of findings. American Journal of Agricultural Economics. 2012;**95**(2):275-281. DOI: 10.1093/ajae/aas037

[47] Vacha L, Karrel J, Kristoufek L, Zilberman D. Time-frequency dynamics of biofuel-fuel-food system. Energy Economics. 2013;**40**(1):233-241. DOI: 10.1016/j.eneco.2013.06.015

[48] Drabik D, Gorter H, Jus DR, Tilmisina GR. The economics of Brazils ethanol-sugar markets, mandates, and tax exemptions. American Journal of Agricultural Economics. 2015;**97**(5):1433-1450. DOI: 10.1093/ajae/aau109

[49] Kristoufek L, Janda K, Zilberman D. Co-movements of ethanol related prices: Evidence from Brazil and the USA. GCB Bioenergy. 2015;**8**(2):346-356. DOI: 10.1111/gcbb.12260

[50] Capitani DHD. Biofuels and food: Can Brazilian ethanol production affect domestic food prices? Economia Aplicada. 2018;**22**(1):141-162. DOI: 10.11606/1980-5330/ea124294

[51] Costa CC, Burnquist HL, Valdes C, Souza MJP. Supply behavior of hydrous ethanol in Brazil. Economia Aplicada. 2015;**19**(4):731-748. DOI: 10.11606/1413-8050/ea137739

[52] Capitani DHD, Cruz Junior JC, Tonin JM. Integration and hedging efficiency between Brazilian and U.S. ethanol markets. Contextus—Revista Contemporânea de Economia e Gestão. 2018;**16**(1):93-117. DOI: 10.19094/contextus.v16i1.1041

[53] Geman H, Ohana S. Time-consistency in managing a commodity portfolio: A dynamics risk measure approach. Journal of Banking & Finance. 2008;**32**(10):1991-2005. DOI: 10.1016/j.jbankfin.2007.05.020

[54] Capitani DHD. Avaliação dos riscos de preços no setor sucroenergético. Revista em Agronegócio e Meio Ambiente. 2016;**9**(3):571-593. DOI: 10.17765/2176-9168.2016v9n3p571-593

[55] Asaftei G. The contribution of product mix versus efficiency and technical change in US banking. Journal of Banking & Finance. 2008;**32**(11):2336-2345. DOI: 10.1016/j.jbankfin.2007.09.026

[56] Burnquist HL. Panorama da safra sucroalcooleira na Região Centro-Sul. Revista Preços Agrícolas. 1999;**14**(1):7-10

[57] Sonoski AAKB, Ribeiro CO. Hedging in the ethanol and sugar

production: Integrating financial and production decisions. Production. 2012;**22**(1):115-123. DOI: 10.1590/S0103-65132011005000069

[58] Capitani DHD, Mattos F, Xavier CEO. Effectiveness marketing strategies and risk measurement in the sugarcane industry. In: Proceedings of 2014 Agricultural and Applied Economics Annual Meeting (AAEA 2014); 27-29 July 2014. Minneapolis: AAEA; 2014. DOI: 10.22004/ag.econ.170270

[59] Brazil. Ministry of Agriculture, Livestock and Food Supply (MAPA). Plano Nacional de Agroenergia (PNA). 2006. Available from: https://agricultura-br.com/. [Accessed: June 04, 2022]

[60] Xavier CEO. Analysis of the Brazilian sugarcane industry efficiency [dissertation]. Piracicaba, Brazil: University of Sao Paulo; 2014. DOI: 10.11606/T.11.2014.tde-03122014-173110

[61] Losekann L, Hallack MCM. Novas energias renováveis no Brasil: desafios e oportunidades. In: De Negri JA, Araújo BCPO, Bacelette R. Desafios da Nação: artigos de apoio. 2nd ed. Brasilia: IPEA; 2018

[62] Kojima. Government response to oil price volatility. 2009. Available from: http://siteresources.worldbank.org/INTOGMC/Resources/10-govt_response-hyperlinked.pdf. [Accessed: June 24, 2022]

[63] Brazilian National Agency for Petroleum, Natural Gas and Biofuels (ANP). Preços. 2022. Available from: http://www.gov.br/anp. [Accessed: January 09, 2022]

[64] Milanez AY, Nyko D, Valente MS, Sousa LC, Bonomi A, Jesus CDF, et al. promessa a realidade: como o etanol celulósico pode revolucionar a indústria da cana-de-açúcar – uma avaliação do potencial competitivo e sugestões de política pública. BNDES Setorial. 2015;**41**:247-294

[65] Eckert CT, Frigo EP, Albrecht LP, Albrecht AJP, Christ D, Santos WG, et al. Maize ethanol production in Brazil: Characteristics and perspectives. Renewable and Sustainable Energy Reviews. 2018;**82**(3):3907-3912. DOI: 10.1016/j.rser.2017.10.082

[66] EPE—Empresa de Pesquisa Energética. Análise de Conjuntura dos Biocombustíveis: ano 2019. 2020. Available from: https://www.epe.gov.br/sites-pt/publicacoes-dados-abertos/publicacoes/PublicacoesArquivos/publicacao-489/Analise_de_Conjuntura_Ano_2019.pdf [Accessed: May 11, 2022]

[67] Li MH, Zhang WD, Hayes D, Arthur R, Yang YT, Wang XD. China's New Nationwide E10 Ethanol Mandate and Its Global Implications. Iowa State University: Center for Agricultural and Rural Development (CARD); 2017. Available from: http://www.card.iastate.edu/ag_policy_review/article/?a=71. [Accessed: March 22, 2022]

[68] Chiaramonti D, Talluri G, Scarlat N, Prussi M. The challenge of forecasting the role of biofuel in EU transport decarbonisation at 2050: A meta-analysis review of published scenarios. Renewable and Sustainable Energy Reviews. 2021;**2021**(139):110715. DOI: 10.1016/j.rser.2021.110715

[69] Lorenzi BR, Andrade THN. O etanol de segunda geração no Brasil: políticas e redes sociotécnicas. Revista Brasileira de Ciências Sociais. 2019;**34**(100):e3410014. DOI: 10.1590/3410014/2019

[70] Grassi MCB, Pereira GAG. Energy-cane and RenovaBio: Brazilian vectors to boost the development of biofuels. Industrial Crops and Products. 2019;**129**:201-205. DOI: 10.1016/j.indcrop.2018.12.006

[71] Brazil. Lei n° 13.576, de 26 de dezembro de 2017. Dispõe sobre a Política Nacional de Biocombustíveis. 2017. Available from: http://www.planalto.gov.br/ccivil_03/_Ato2015-2018/2017/Lei/L13576.htm. [Accessed: June 22, 2022]

[72] Salina FH, Almeida IA, Bittencourt FR. RenovaBio opportunities and biofuels outlook in Brazil. In: Sayigh A, editor. Renewable Energy and Sustainable Buildings. Innovative Renewable Energy. Cham: Springer; 2020. DOI: 10.1007/978-3-030-18488-9_30

[73] Lazaro LLB, Thomaz LF. A Participação de stakeholders na formulação da política brasileira de biocombustíveis (RenovaBio). Ambiente e Sociedade. 2021;**24**:e00562. DOI: 10.1590/1809-4422asoc20200056r2vu2021L4DE

Section 3

Sugarcane Cultivation and Response to Abiotic Stresses

Chapter 6

The Social Role of Sugar Cane Cultivation in the Fertile Gharb Region of Morocco

Mounia Elhaddadi

Abstract

A special report had been achieved in the GHARB region of Morocco; known as a large plain with abundant pure water sources especially rivers such as the SEBOU permanent river and others, very heavy clay soil sometimes so dark, and a Mediterranean climate; those conditions ameliorate the rate of agricultural activities including unique ones such as Sugar cane/Rice/Tobacco, etc. Sugar cane plays a crucial role in the population's lifestyle's amelioration and stability by improving their social and economic conditions besides sustainable development. The report that designed four categories of population afterward chose one best representative of each category to interrogate. The analysis of the statements showed similarities with universal values harvested from sugar cane cultivation in addition to unique ones related to the unicity of the geographic, cultural, and demographic data.

Keywords: sugar cane, Gharb, Cosumar, agriculture, Morocco, peasant, sustainable development

1. Introduction

Sugar cane is a special plant in a large zone of the southern EARTH, relating to its nutritional value especially carbohydrates, and its socioeconomical role that covers securing jobs and using its products as alternative solutions for construction, energy even daily activities material.

We used the exploratory method to discover the development and sustainability that the sugar cane cultivation brings to the lives of people in the GHARB region of Morocco, based on our own method we name: targeted reporting; the method that categorizes the population according to its potential impact on the subject and chooses the best representative who has knowledge and expertise that touch the top. Four categories were detected: ancient peasants-ancient workers-History witnesses-New workers. Four persons to allocate an interesting amount of affirmations leading to an adaptive briefing on the topic.

How was the selection of participants managed ? And what is the size of the statement added by their testimony ? Is sugar cane cultivation considered a major contributor to moving the wheel of the economy and sustainable development ?

On one side, Sugar cane cultivation in GHARB region of Morocco has some unicity and special characteristics such as spreading new lifestyles and the multiple uses of its residues and leftovers; on the other side, it ensures the same needs it ensures in many other world regions as jobs, economic network activity, industrial development and securing the demand on sugar.

2. Sugar cane in the world: crop growth period varies by region

Most sugar cane is grown in subtropical and tropical regions, especially in Brazil and southern of Asia, before spreading to other world regions such as North Africa, and it is the principal source for extracting sugar after it comes sugar beet. Its cultivation needs fertile land and an important quantity of water and remains in the ground for a whole year, and the sugar factories are in the middle of the cane farms. The leftover cane is used in some para-agricultural activities such as the manufacture of alcohol famous in Brazil [1].

The people of the South Pacific islands were among the first to grow sugar cane to use its stems as stick, in a mixture of sand, clay, and silt particles with organic matter at a temperature of 20°C. To 7 meters long, it has long, sword-shaped leaves and a number of buds through which the crop propagates, with a density of 10000–25000 sugar canes per hectare [2].

In Australia, it takes 15 months to grow, while it takes 9 months in Louisiana in the USA. The sugar cane crop is harvested three times in dry and cold seasons that last for 6 months so that three economic crops are harvested from one original cultivation and after the end of the economic crops, all roots are uprooted, and the field is replanted again [2].

The soil must be fertilized with nitrogen, phosphorous, and potassium fertilizers according to the nature of the soil and climatic conditions, as well as the use of chemical herbicides to eliminate weeds and damaged insects. Brown sugar is extracted from a brown liquid inside the hull of the sugar cane, and then the brown liquid is repeated to produce white sugar [3] .

Sugar crops have reached an important level of development, mainly due to the mechanization of crops and the introduction of new production technologies, noting that the combination of factors contributes to improving the raw incomes of producers by more than 10% for these two types of crops. The use of good seeds, early maintenance, irrigation, and appropriate climatic conditions contributed to the achievement of diabetic beet cultivation program and improved the status of sugar crops [4].

3. Sugar cane cultivation in Morocco (Region of Gharb Cherarda Benihessen)

Agricultural policy is reviewed in historical perspective, to show that the liberalization process, which was proposed in the framework of structural adjustment reforms, ran contrary to the agricultural development strategy followed by Morocco since Independence [5].

In Morocco, farmers benefited, for sugar cane production, from a subsidy of up to 6000 dirhams granted by the state, in order to make the new cultivated areas succeed, in addition to a grant of 3000 dirhams distributed between the state and the Moroccan Interprofessional Federation of Sugar to encourage early planting.

Figure 1.
Sugar cane season Morocco 2018–2019.

Given that the success of sugar plantations remains dependent on irrigation, the state, adds the official, launched the national program for saving irrigation water to encourage farmers to use techniques that contribute to rationalizing irrigation water (the drip irrigation system), noting that support rates range between 80 and 100%, depending on the area allocated [6]. To conserve the freshwater with which the sugar cane is irrigated, pivot and canal irrigation switched to drip, so the farmer consumes about 14%0 less water than before (**Figure 1**).

The introduction of sugar cane in the region contributed to improving the individual income, as the economic situation of the families, who were living on nomadic livestock and cultivation of wheat and barley, has been improved

This cultivation activity suffered from intense competition from the other regions of the Kingdom, which are on the throne of grain production. The Gharb region also experiences more floods every year due to its low topography and the presence of the large and ever-flowing Sebou River, on which a dam was built in the outskirts of Fez. However, this infrastructure causes floods with the high water level during the winter season.

4. The role of cosumar sugar company in the Gharb's Region's sugar cane cultivation improvement

Cosumar Sugar Company "COSUMAR" is the first of its kind in the sugar industry in the Kingdom of Morocco, with a production capacity of more than 800,000 tons of white sugar annually. It is a public company listed since 1985 on the Casablanca Stock Exchange and was established in April 1929. The history of Cosumar merges with the beginning of the industrialization of the production of white sugar in Morocco. Formerly of artisanal manufacture, it was in 1929 that the first sugar refinery was created in Casablanca in the Roches Noires district. Let's discover the key dates that mark the history of this great Moroccan company [7].

As a major sugar Moroccan company, it was a given in recognition of the role played by the COSUMAR Group as an aggregator in the sugar chain. Indeed, COSUMAR is involved in all levels of the sugar chain, especially at the level of agricultural activity, with the aim of achieving a common goal: the development, competitiveness, and sustainability of the national sugar chain by encouraging the cultivation of cane Sugar and sugar beets. As COSUMAR, an investor in the food industry and a responsible and solidarity group, it keeps pace with its peasant partners at the financial, technical, and social levels, and the collection agreements that it brought and established by the Green Morocco Plan aim to strengthen the relationship in a win-win manner with its peasant partners. This aggregation role revolves around many activities, including:

- Funding support and technical and human supervision for sugar plant producers in order to obtain the best returns that help to achieve a significant increase in incomes.

- Ensure the transfer of the agricultural crop to the processing plants.

- Ensuring complete transparency in performance thanks to the modernization of the raw materials reception centers.

- Product purchase guarantee.

- Creation of a solidarity fund for the protection of farmers in the case of low yields due to climatic difficulties, within the framework of the Moroccan Interprofessional Federation of Sugar.

- Social support for peasant families [8].

5. History of sugar cane cultivation in Morocco

The sugar industry is not a novelty in the life of Morocco. Rather, it had in the past, as it may have in the future, a clear impact on the economy of our country, and even in its internal policy, as well as in its relations with some countries. However, drunkenness in ancient Morocco, unfortunately, did not receive the studies it deserves, and Moroccan history books have preserved for us only brief references about this vital substance. Perhaps it is useful to review in this hurry some aspects of the subject, so we know where sugar cane was grown, and where it was made, and then the relationship of sugar to the policy of the state at the time (**Figure 2**) [9].

The first historians who spoke about Morocco in the form of Abu Hanifa al-Dinuri, Ibn Hawqal, al-Bakri, Ibn Khaldun, and others, all of them indicated that this agriculture was very prosperous in Morocco, and these references are almost nonexistent in what was written about Morocco before the advent of Islam as well as after the year 1615, which made us. We believe that the age of this prosperity began in Morocco with the Islamic conquest and that sugar cane was brought by the Arabs, just as they brought other crops such as orange trees. Consequently, sugar cane has lived in Morocco for a period equivalent to eight centuries, i.e., in the year 895 AD to the year 1615 AD. As for the places where this cultivation was known in Morocco, it is in the south of Morocco in particular [9].

We may be surprised if we learn that this cultivation was also in Mount Tanmal, the cradle of the Almohads, and near the city of Salé, as well as around the city of Ceuta. The ancients call the south of Morocco the Souss region, and by this name, they mean the area between Heri Chichaoua and the Draa Valley. It includes the

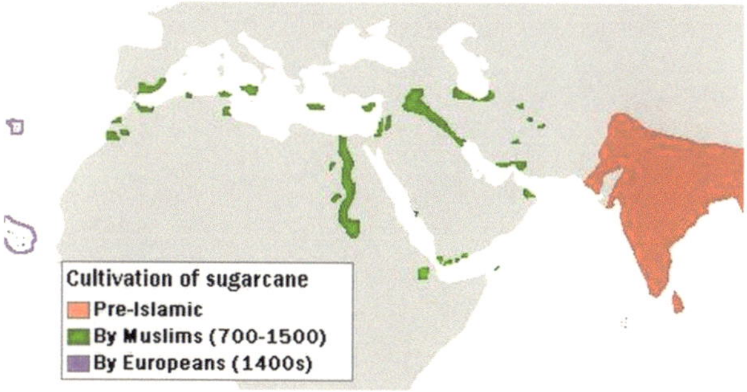

Figure 2.
Map showing sugar cane India as the origin of the westward spread, followed by small areas in Africa and then smaller areas on atlantic islands west of Africa, https://stringfixer.com/ar, 14/8/2022.Morocco situation in the far Northwest of Africa with green spots showing sugar cane activities.

mountainous Atlantic region adjacent to the sea, the Souss Valley, and the Little Atlas to the Draa Valley. This region was one of the most fertile regions of Morocco, if not the most fertile at all. The Moroccan travelers, nor historians, in describing a region of Morocco, did not mention this region. It was one of the most densely populated areas.

Al-Idrisi, in turn, pointed out that sugar cane is found in Souss; especially in the district of Taroudant; in Ceuta and Tibnmel. He said about the region of Souss: "The country of Souss has many villages, and their architecture is connected to each other, and there are great fruits of different types, and many types, such as walnuts, figs, grapes, apricots, pomegranates, peaches, apples, and reeds." Sugar, which is not on the ground of the earth like it in length and breadth, sweetness, and abundance of water, works in the country of licorice from the sugar attributed to it that pervades most of the earth, and it is equal to Sulaymani sugar and tabarzad, rather it heals all types of sugar in goodness and purity. Regarding Jabal (mountain) Darn, Al-Idrisi says: "...that the people of this mountain do not sell it among themselves (i.e. sugar cane) nor buy it because of its abundance." The owner of the insight also indicates that Taroudant, which is a large village, is the richest country in God with sugar cane, and that its mills are more numerous, carrying sugar from it to the rest of the countries of the Maghreb. After this pause on the territory of Souss, and after introducing it, we can identify the areas where sugar cane was cultivated and divide them into three groups:

1. The North Group, including Tangiers, Ceuta and Salé.

2. Al Haouz Group, including Chichaoua, Essaouira, the Qasab River Basin, Wadi Nafis, and Zaouia Sidi Chiker.

3. Souss group according to the aforementioned meaning. It is well known that this sugar cane cultivation will not be of any benefit as long as it is not manufactured and the industry needs a motive power, so from where did the technicians in this age obtain the motive power to accomplish this important work?

6. The impact of sugar cane cultivation in Gharb's population life

This research was carried out as a reportage with farmers, workers, and producers who belong to the western region of Morocco, where sugar cane cultivation flourishes.

Figure 3.
2015, Kingdom of Morocco map, GHARB CHRARDA region in REd.

The nomadic life changed and settled after the settlement of sugar cane cultivation in the region, where people settled for its production, neglecting the livestock breeding that used to occupy most of their activity. In order to rationalize the use of water and determine its quantity, center pivot and drip irrigation were introduced, which made the used amount reduced by half. Thus, the previously wasted water was preserved, especially the arable water, because the groundwater is characterized by a high salinity rate, which is difficult to exploit in agriculture. Also, the river Sebou is not close to all the farms and the horizontal canals that carry its water, as it was built during the French colonization of Morocco between 1912 and 1956. The vast area of Gharb Chrarda does not feed Beni Hessen (**Figure 3**).

7. Method and discussion

To achieve our method of questioning that had the following purpose: highlight the social role of sugar cane cultivation in the fertile Gharb region of Morocco was needful to divide GHARB area's population to four categories responding to the following directed questions:

- Who are the main players in the activities related to sugar cane cultivation?
- Was there a reversal impact between the enhancement of this population lifestyle and the introducing of sugar cane cultivation in the area?
- What are the sectors of the population life that were affected by this agricultural activity?

Responding to the questions above we have detected four interesting categories of people that are in a close and mutual attachment with sugar cane cultivation:

- New workers
- Peasants
- Ancient workers
- History witnesses

To find the perfect person for each category, we stand in touch for about a month with the population to achieve our goal. So four people were selected and interviewed first in their zone of living; second, we have completed the operation via WhatsApp application for instant chats.

8. Description of the operation

See **Table 1**.

9. Details and types of participants' main contributions

Social:

- Replacement of The mud houses and tents by cement houses
- Infrasructure
- Settelment of many villages
- Demographic growth
- Prosperity of the social interaction

Participant	Statute	Importance of the statement	Types of the testaments contributions	Category
MISS. Heba Al-Azeeb « student/worker/resident »	Student	Important	• Economic • Geographic • Social	New workers
Mr. Houssain Maymouni « Farmer/resident »	Sugar cane Farmer	Important	• Professional • Historical	Peasant
Mr. Mohammed Alhafian «retired/resident »	Retired worker	Verry impotant	• Professional	Ancient workers
Hajja Ruqayyah Limouna « resident »	Elder of a Gharb familly	Verry important	• Social • Cultural • Historical	History witnesses

Table 1.
Description of the sample of the operation and their testaments.

Economic:

- Securing jobs
- Economic boost of the region
- Commercial popularity
- Connecting regions economic network

Professional:

- COSUMAR as the mother factory of sugar raffination in the region
- The discription of the cultivation cycle
- Planting technics
- Preparation of the sticks for squeezing and stocking

10. Summary of the operation

MISS. Heba Al-Azeeb. The girl stated that sugar cane is cultivated in the direction of the Belkassiri town from Kenitra city, specifically in the lands or fields where mud is aboundant in the soil formulat. She added that the work is carried out on a random basis and not in an operating system without specified timing. The work in the fields starts from about 10 in the morning until sunset. The upper limbs of sticks are cut, leaves are removed, and it is peeled with knives by a group of female workers. The male laborer is responsible for planting and watering the reeds; and it is thrown on the ground; a solution of javel and water is thrown on it to facilitate the process of cleaning by another group of workers after rubbing and washing it well; another group performs. By collecting it in the form of groups of packages and making it stand and pouring water on it, then the owner comes to load it on tracks and distributes it to the mills where it is squeezed; in the case of free planting; or by COSUMAR company in the case of the contaractual planting. And this is the case throughout the year, the girl says, because the product is always in demand and many customers buy it. An expert peasant like **Mr. Houssain Maymouni** gifted a professional affirmation about the course of choosing and planting technics in GHARB region of Morocco, pointing that the most structured planting plan is the one that held by a contract with COSUMAR company. COSUMAR requires an area of the field to be 1 hectare owned in the name of the beneficiary as the reason of a contract redaction. According to **Mr. Mohammed Alhafian,** the ancient worker one contract for each hectare of land, he proceeded: factory committees come to examine the land and water import possibilities before redacting the contract. They also investigate that the beneficiary is able to complete the task of planting and taking care of this planting and sugar cane germination according to the conditions agreed upon. This contract must be approved in the official departments between the factory and the farmer (**Figure 4**).

By signing a contract the factory releases the planting seedlings according to conditions and controls. With regard to irrigation, the farmer; after the contract with the

Figure 4.
Slogan edited by COSUMAR in 2018 to organize a yearly social cooperation compain, www.cosumar.co.ma 12/08/2022.

sugar factory is issued; writes a request to the Office of Water and Electricity, so that in that request he shows the contract that is required of him to follow its provisions for cultivation, and determines its place, then writes the application and signs it and puts it under the officials of the Office. Water expenses are paid every year after the harvest. Both participants described the preparation of land to start by aeration and mashing using fertilizers at the farmer's expense and with his own effort. With regard to agriculture in Morocco, the state specified that between tow seedlings there should be a distance of one and a half meters. The sapling is struck at the top to open it, then another sapling is placed on top of it, and this unit or this compound is planted. After the process is completed, fertilizers are also added on top of the planting. The previous operations include a large workforce mostly from the region that increases social cooperation, but there are disadvantages too: The by-products produced during processing and refining procedures are discussed along with the ways that they affect the environment. The wastes produced by sugar mills are also shown to result in heavy metal contamination and parasitic and bacterial infections [10]. **Hajja Ruqayyah Limouna**, an 84-year-old lady, conforms that disadvantages such as the allergy she had don't cover the huge benifict, so she added that mud houses, huts, and tents were replaced by cement houses, roads constructed, and settled villages appeared after the construction. The SWEET GOLD increased celebrations such as weddings, engagements, national and religious holidays, to become more resonant and very fascinating. Stability and organizing the annual income had a great impact on food habits, meals and ingredients have become diversified, the prosperity of livestock breeding and selling and the increase in demand for it (**Figure 5**).

On the environmental level, **Mr. Maimouni** says that the atmosphere has become more humid and the high temperatures have decreased, as well as the disappearance of sand and dust storms because of the sugar cane fields that act as pumps for water vapor, as well as natural barriers.

Also, the places of spread of insects that abound during the period of heat and humidity have been redistributed. They find refuge in his fields, and they are reduced in the places of the population. Also, rodents have multiplied and their number is equal to the number of poisonous reptiles that provide food for them.

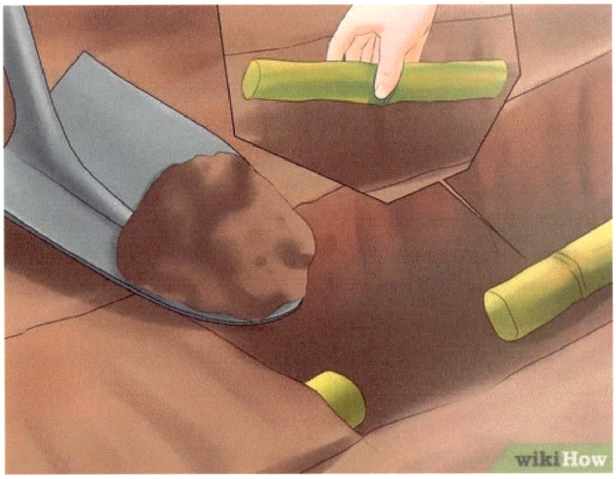

Figure 5.
WIKIHOW [11], cultivation of sugar cane, www.wikihow.com 15/8/2022.

Heba Al-Azaib continued citing that sugar cane fields are a haven for nesting and hatching many poultry, such as ducks and turkeys, as well as wild chickens and turkeys, which are birds that have been domesticated and bred since ancient times, but they did not retreat and remained with these fields a basic activity The region and another source of profit and food.

As for **Muhammed Al-Hafyan**, he said that the sugar cane residues in the sugar refinery were burned, but recently it has been sent to centers for thermal energy extraction experiments, while in the village the older cane is used to feed livestock and to bake bread inside the clay ovens, and these residues are recovered after drying them are mashed as a support for the silt that covered the walls of mud houses, huts, and traditional stables, as well as poultry and livestock houses.

When sugar cane is grown, they exploit its product for a period of about 3–4 years, after which they extract it from seed and plant a new one to give a more profitable production. Method of work: Its upper limbs are cut, leaves are removed, and it is peeled with knives by a group of female workers, and it is thrown on the ground and a solution of javel and water is thrown on it to facilitate the process of washing it by another group of workers after rubbing and washing it well, another group performs. By collecting it in the form of groups of packages and making it stand and pouring water on it to complete its washing, then its owner comes to load it on his cart and distributes it to the owners of the mills where it is squeezed. And this is the case throughout the year, the girl says, because the product is always in demand and everyone buys it. As for the hardship of work, of course it must be, especially since the girls work using the Javel (chlore) solution, and the cane must make its scratches, which sometimes have severe injuries, knowing that everyone knows that this is the case, because the operator has made them aware of that at first. Wages are paid weekly every Friday. He states that if the area of the field is hectares, then they plow with the big plow, and this plowing process is followed by other stages, such as the process of cherub and combing the field well, forming compact lines, and they put compost in it, they call it ~ dust ~ and then they plant the reeds after selection, examination and good and carefully purification, that is, the branch that will be planted is purified from any dirt or rotten leaves, and they leave one pure leaf to cover what they call

Figure 6.
Youssef [12], Cosumar logo and its promotion for its support thouward population's creation and development.

the eye of the branch, which they give very carefully in the process of planting, after which they put the compost. Mr. Maymouni adds that Type 65 is a good variety with good quality and juice production. The stalk of this type is white, with a distance of 20 centimeters between its rings, and this type is good. He also adds that the types of cane are numbered. Before proceeding to the germination of sugar cane, a contract is drawn up with the sugar factory associated with the Gharb Chararda Benihessen region. Planting and when the factory or factory committees will come to harvest, and this contract is issued for each hectare of land. The only free farming, farming for sale to sugar cane juice makers, is very small-scale farming, the farmer said. Preparing the land is at the expense of the farmer, as he plows and aerates the soil many times using a mechanical plow and also using daily farmers and also using a traditional plow with the help of an animal, in order to prepare the soil for the cultivation of sugar cane. The farmer contributes to the harvest, but for carrying, cleaning, and making use of the sugar cane, this costs the factory, starting from carrying it from the harvest area to the factory. Farmers do not start preparing the land for cultivation by the process of preparing the land is also done with regard to irrigation, the farmer Madame (**Figure 6**).

11. Conclusion

The female farmer choses sugar cane as an activity, which is an important percentage of the families. The mud houses and tents were replaced by cement houses, and the activity of cutting the road, which polluted their reputation. There are three religious worships: Eid al-Fitr, Eid al-Adha, and the birthday of the Prophet Eid Al-Fitr comes after the end of Ramadan, and it is the first day of breaking the fast, and its first date is in the month of Shawwal according to the Hijri calendar. On this Eid, a special prayer is held in the morning, followed by a varied and carefully prepared Iftar feast. People wear the most beautiful clothes and visit the people of the dead in the cemetery. Stability in which there is life, Abdul-Fitr is also linked to giving out an obligatory charity called zakat, and the people of zakat are the poor, and your presence in a stable place periodically gives charity to the same poor, you. There is a purely Moroccan activity, which is the celebration of the end of the agricultural seasons, which is a festival held in a public place in the wealthy and lasts for a week in which tents are built and all foodstuffs and tools are sold. The stealth race of a group of cavalry with simultaneous firing of gunpowder from their rifles. Each village has its own troop of horses and a herd of purebred Berber horses. The Western region has become competitive with other regions after it has known agricultural growth and economic boost since the 1970s of the twentieth century, and these celebrations are still being held so far, and the Western cavalry teams are still occupying advanced positions. The stable social and economic situation created by the sugar cane cultivation introduction made the need for infrastructure urgent, so roads were supplied and the number of primary, middle, and secondary schools increased as well as the spread of police stations, the royal gendarmerie points, post offices, municipal building; a decrease in rural immigration toward urban compounds were observed; besides, young population have kept living in the rural zone so agricultural activities remain in its vital cycle. Sugar cane growers and ivestissors participate to the establishing of a sustainable life because they often associate with them day laborers, agricultural technicians, farmers, mechanics, sellers of agricultural tools and fertilizers, drivers, and food investissors. This network, created due to this agricultural activity continuum, creates an acceptable to moderate living situation for thousands of families. Improving wages and intensifying efforts to protect the environment from the effects of the remnants of sugar refining activities, in addition to establishing associations and cooperatives for medical treatment and supervision, has become a requirement of activists in the sector [13–19].

Acknowledgements

Special thanks to:

- Miss. Heba laazib "student from OULED SLAMA rural commune"
- Mr. Houssain Maymouni « Farmer/resident of MOGHRANE rural commune »
- Mr. Mohammed Elhafyane « retired and resident of SID LKAMEL rural commune »
- Hajja Ruqayyah Limouna « Elder/resident of OULED SLAMA rural commune»

On the environmental level, Mr. Maimouni says that the atmosphere has become more humid and the high temperatures have decreased, as well as the disappearance of sand and dust storms because of the sugar cane fields that act as pumps for water vapor, as well as natural barriers. Also, the places of spread of insects that abound during the period of heat and humidity have been redistributed. They find refuge in his fields, and they are reduced in the places of the population. Also, rodents have multiplied and their number is equal to the number of poisonous reptiles that provide food for them.

Author details

Mounia Elhaddadi
Faculty of Sciences, Cognitive-Behavioral Neurosciences and Applied Nutrition Unit, Department of Biology, Biology and Health Laboratory, Ibn Tofail University, Kenitra, Kenitra, Morocco

*Address all correspondence to: elhaddadi24@yahoo.fr

References

[1] Anonyme. Plants & fungi: *Saccharum officinarum* (sugar cane). *Royal Botanical Gardens*, Kew. 2021b. Archived from the original on 4 June 2012

[2] Yamane T. *Encyclopedia Britannica* "sugarcane". 26 Dec. 2019. https://www.britannica.com/plant/sugarcane. [visited 12/8/2022]

[3] Gatiboni L. Soils and plant nutrients. In: Moore KA, Bradley LK, editors. North Carolina Extension Gardener Handbook. Raleigh, NC: NC State Extension; 2018

[4] Anonyme. Sustainable Agricultural Mechanization. FAO; 2022d

[5] Kydd J, Thoyer S. Structural Adjustment and Moroccan Agriculture: An Assessment of the Reforms in the Sugar and Cereal Sectors. Paris: OECD Development Centre; 1992

[6] Anonyme. *Guide investisseur-AR. PDF*, Investor's Guide to the Agricultural Sector in Morocco. 2018. https://www.ada.gov.ma/. [visited 12/8/2022]

[7] Anonyme. The Ministry of Awqaf and Islamic Affairs. Sugar in old Morocco, *Daawat Alhak Magazine*, No 111,112. 2022b. https://www.habous.gov.ma/. [visited in 12/05/2022]

[8] Anonyme. *COSUMAR*, sugarcane role in peoples life. 2022a. https://www.cosumar.co.ma/ar. [visited in 20/05/2022]

[9] Zaimeche S. *Foundation for science technology and civilisation, Microsoft Word - Morocco June 03Edit4PDF.doc*, Morocco as a great centre of Islamic science and civilization. 2004

[10] Travis G-R. An overview of sugar culture in Morocco, particularly within a Berber community in Rastabouda, *SECHERESSE*. 2007. http://www.secheresse.info/. [visited 11/8/2022]

[11] Anonyme. *WIKIHOW*, cultivation of sugarcane. 2022f. www.wikihow.com. [visited 15/8/2022] (figures of planting)

[12] Youssef. *LESITEINFO*, Cosumar logo and its promotion for its support thouward population's creation and development. 2022. www.lesiteinfo.com. [visited 12/8/22] (photo)

[13] Anonyme. *Almokawil magazine*, meet cosumar of sugar production. 2022c. https://www.almokawil.com/. [visited 03/06/2022]

[14] Bailey LH, Bailey EZ. Hortus Third: A Concise Dictionary of Plants Cultivated in the United States and Canada. New York: MacMillan Publishing Company; 1976

[15] Anonyme. *Yummi Magazine*, sugarcane benificts. 2021a. https://yummy.layalina.com/. [visited 5/5/2022]

[16] Anonyme. *Rapport-annuel-Cosumar-2019-Web-en_reduce.pdf*, 90 years of shared values. 2019a. https://www.cosumar.co.ma/. [visited 14/8/2022]

[17] Anonyme. *stringfixer*, cutivation of sugarcane. 2022e. https://stringfixer.com/ar. [visited 12/8/2022] (map)

[18] Anonyme. *Algharbchrarda blogspot*, Algharb chrarda region. 2015. http://algharbchrarda.blogspot.com/. [visited 14/8/2022] (map)

[19] Anonyme. *Sugar planting season 2018-2019 in the West*: A record production of more than 1.6 million tons of sugar beet and sugar cane is expected, MAP ECOLOGY, 2019b. https://mapecology.ma/ar [visited 11/82022] (photo)

Chapter 7

Sugarcane Response and Its Related Gene Expression under Water Stress Condition

Abhisek Shrestha, Bharti Thapa and Ganga Dulal

Abstract

This review paper is to study the different responses expressed by the sugarcane when exposed to water stress conditions, that is, waterlogging and drought. Water stress is one of abiotic stress affecting sugarcane productivity and the development of water-use efficiency and the morphological character get varies with genotypes, duration and intensity of stress and types of tissue damage and expression of variable patterns of a gene that makes a high degree of complexity on sugarcane under water stress condition. Since, there is little stepping towards sugarcane crops coming from genetics, agronomics, and molecular biology. These studies provided the framework for researching the morphological basis of genetic variation and mycorrhizal colonization in water stress tolerance and yield improvement under water-limited conditions.

Keywords: sugarcane, gene, yield, water stress, complexity

1. Introduction

Sugarcane (*Saccharum officinarum* L.) is a unique crop that can accumulate high levels of sugar and is also a commercially viable source of bioelectricity and second-generation bio-ethanol. It provides over 76% of the sugar for human consumption from approximately 27.12 million hectares area with the production of 1900 million metric tons around the globe [1]. Brazil is the leading country in terms of sugarcane area and production followed by India and China [1]. Sugarcane provides juice, which is used for making white sugar, jaggery (gur), and many by-products like bagasse and molasses. Sugar and its fermented products are very important in making and preserving various kinds of medicines like syrups, liquids; capsules, etc. Bagasse is used as a fuel, for the production of fiberboard, papers, plastics, and furfural. Molasses is used in distilleries for the manufacture of ethyl alcohol, butyl alcohol, citric acid, ethanol, etc. Rum is the best potable spirit made from molasses. Molasses is also used as an additive to feed livestock. Green tops of cane are a good source of fodder for cattle. Press-mud, the remains after juice clarification is good manure. Steam produced during the boiling of the juice is used to generate electrical power. Despite its importance in production and productivity sugarcane is affected by various biotic and abiotic stresses. On the other hand, no use of improved and recommended varieties also causes yield decline, among the abiotic stresses Water stress Is one of abiotic stress abiotic stress affecting sugarcane productivity. Therefore, the development of water-use efficient and water-stress-tolerant cultivars is imperative for major sugar-producing countries and industries. The susceptibility of sugarcane to water stress is greatly exposed in tillering and stem elongation phases with both leaf and stem growth [2] but has a positive effect in the maturation phase with sucrose content. The morphological character get varies with genotypes, duration, intensity of stress, and types of tissue damaged [3]. However, reported large potentially exploitable genetic variation for cane yield, responses like leaf rolling, inhibition of stalk and leaf growth, leaf senescence and reduced leaf area, and root development pattern [4]. Knowledge about stress biology in other crops increases coming from genetics, agronomic and molecular biology but still, there is little stepping towards sugarcane crops.

The use of arbuscular mycorrhizal fungi (AMF) has significant importance for more efficient crop development and the use of fertilization, especially in the early stage of development [5], and helps in the absorption of soil nutrients like phosphorus and plant resistance to biotic and abiotic stress. Abdel-Fattah & Asrar [6] reported that its association increases tolerance to water stress, contributing to aggregate stability, increase in soil aeration and subsequent availability [7], the reason behind, plants maintain higher stomatal conductance during water deficit periods and, in return, higher diffusive dependence. However, studies on interaction between AMF and sugarcane are scarce, and need to develop studies of AMF in sugarcane for different water stress conditions in lab conditions to explore the importance to cope with different stress situations.

Plant species go under different alternations in antioxidant enzyme activities. Drought tolerant crops showed the up-regulation of SOD and CAT activity as a general mechanism, where there is a remarkable correlation between antioxidant and proline content that signifies the specific traits for future identification of drought tolerance [8] that contributes protection against oxidative damage [9]. These enzymes were further affected by plant species, cultivars, and stress intensity and duration.

From the perspective of molecular studies, a wide range of gene are explicitly recognized that are expressed in sugarcane under water stress condition [10] that encodes putative chaperones dehydrin (DEH) and late embryogenic (LEA) protein, enzymes involved for metabolism and oxidation of proline [10]. There is an expression of variable patterns of gene that makes a high degree of complexity on sugarcane under water stress conditions. On other hand, signal transduction pathways lead to stress response due to plant hormones like ABA play an important role in signaling and gene regulation. Under water deficit, ABA concentration increases and makes stomata functional to protect against rapid desiccation [11]. Use of beneficial nitrogen fixing bacteria and gibberellic acid (GA_3) involves in gene expression and regulates the metabolic process as a function of sugar signaling and antioxidative enzymes [12].

Study of the morphological and stress response helps us to develop the concept for future breeding programs and selection of the water stress tolerant genotypes. In the present context, there are limited varieties named Jitpur series, that does not resolve the problem of water stress condition, so further study should be conducted for studying the response to water stress and selection of genotypes under different level of stress condition. So, this research should be conducted for studying the morphological and stress-responsive character of different genotypes for better crop improvement and to find a suitable variety for water stress conditions.

The knowledge of response and genetic variation of sugarcane to water stress under field conditions is relatively limited [13] and restricted in a small number of studies. Most studies on water deficits in sugarcane have focused on irrigation management practices [14, 15]. A number of physiological investigations conducted under laboratory or glasshouse conditions together with modeling studies have advanced the knowledge of water relations, stomatal functions and osmoregulation, carbon assimilation, and dry matter partitioning in water-stressed sugarcane [2, 3, 16–20]. Sugarcane breeding program has access to a broad range of parental germplasm and cultivars developed more than 100 years of breeding with selection, which derives from wild cane are relatively less developed and selected.

Despite the other major consequences of climate change, it is predicted that sugarcane yield will increase with increasing average minimum and maximum temperatures in winter, and increasing maximum temperatures in the rainy seasons [21]. Recent Nepalese climatic projections support the above assumptions and suggest that these conditions are likely. However, recent studies on the impacts of climate change on sugarcane [22] indicate that climate change-induced drought is one of the most significant challenges for sugarcane production. Moreover, the crop is more able to withstand natural disasters such as flood than rice and/or other cereal crops. Regardless of the high resilience to natural disasters, less than half of the potential of sugarcane production is currently realized [23]. The morphological basis of genetic variation and mycorrhizal colonization under different water stress situations were studied.

2. Morphological responsiveness of sugarcane under water stress condition

Morphological character is the first character seen when sugarcane gets exposed to water stress conditions. On exposure to drought conditions, there is a reduction in tillering, leaves discoloration, rolling of leaves and leaves folding and shredding, reduced leaf area due to narrow leaves, and decrease in lipid peroxidation. Under the waterlogged conditions, the sugarcane showed pipping in stalks and roots, development of adventitious roots, and the presence of aerenchyma tissue without which sugarcane cannot survive in water-logged condition.

Sugarcane, an important source of sugar and ethanol, demands relatively high water and is highly sensitive to water deficit [24]. The use of one Ml of irrigation produces 8–12 ton cane [25] and water deficit can lead to 60% productivity losses [26]. For this reason, production areas are concentrated in regions with favorable rain regimes for sugarcane growth and development [27], while in other areas crop production requires supplemental or full irrigation [28]. The increasing incidence, duration, and intensity of severe water deficit have promoted large sugarcane crop improvement programs for water use efficiency and water stress tolerant varieties that make impetus to develop biotechnological strategies.

From the perspective of growth stages and water management practices, susceptibility is more in tillering stage ad stem elongation phases [2] with both stem and leaf growth being more affected than other organs. The morphological and physiological responses of sugarcane plants get varies with genotypes and duration and intensity of stress and type of tissue damaged and substantially decreased both cane and cane yield. There, as reported, has a large potentially exploitable genetic variation for cane and cane yield under stress conditions. The most common water stress responses in sugarcane are leaf rolling, stomatal closure, inhibition of stalk and leaf growth, leaf senescence, and reduced leaf area [2]. Besides cell division and elongation get interrupted, so stem and leaf elongation and root development are more seriously affected during growth processes [4, 29].

There is an interesting fact that increased levels of sugar, such as trehalose, can help plants to cope with water deficit, reducing the damage to cell membrane [30].

3. Mycorrhizal colonization variance in water stress condition

Water stress strongly affects mycorrhizal development in roots and soils. The short-term soil drought did not appear to favor or discourage root AMF colonization and longer-term soil drought decreased AMF colonization. So, there are increased levels of root AMF colonization in response to drought stress than in decreased levels, which is related to reductions in plant P levels.

The growth of AMF spores varies with soil moisture available, which alters spore behaviors. Spore germination is favored n soil at or above field capacity but is decreased with decreasing soil water potentials below field capacity [31]. AM-stimulated plant growth enhancement may be more important in the host plant under drought stress than under watered conditions. In addition, helps in increase of P nutrition, water uptake by hyphae, and increase of root length [32].

The survivability of plants under stress conditions is strongly enhanced with AMF inoculation. This may be due to a more effective root architecture for water and nutritional absorption and the developed external hyphae in the soil [33]. Mycorrhizal plants would recover more quickly from wilting than non-mycorrhizal plants after drought recovery [34]. In addition, leaf morphological adaptation, and root morphological adaptation are other strategies for mycorrhization under drought stress. This adaptation provides more exploration of soil volume to absorb water and nutrients from the soil [35], thereby potentially enhancing the drought tolerance of the host plant. Mycorrhizal hyphae help in the uptake and transport of water from bulk soil to the host plant and also considerable quantities of phosphates and nitrogen to the plant from soil zones but yet no significant evidence of transfer of water from hyphae to plants. In sum, AMF-enhanced nutrient absorption is an important physiological mechanism in drought tolerance of the host plant caused by mycorrhization.

4. Genetic variability of sugarcane under water stress condition

Crop growth under water stress has been frequently and usefully conceptualized as the product between the components of total water use and water use efficiency. The high genetic correlation between water levels is likely traits for contributing to increased levels of component under fully irrigated and contributed to high yield under water stress. Genetic gains accomplished from long-term selection in breeding programs conducted under relatively non-stressed situations extend large degrees under water stress.

In relation to water use efficiency, high photosynthetic capacity improves rates of crop growth under non-water-limiting conditions and also improves water use efficiency through reducing leaf internal CO_2 levels, thus, contributing to improved performance also under water-stress conditions. Similar to water use, high radiation use efficiency and fast canopy development under non-limiting conditions contribute to increases above- and below-ground (root) growth. Genotypes with larger canopies and root systems have a greater rate of water use and efficiency and growth which may persist for some time even as stress develops, because larger canopies and above-ground growth support greater root growth, and vice versa. However, when soil water is depleted to a point where extensive root systems cannot provide a benefit (because there is little or no water remaining), the initially vigorous genotypes with relatively high water use would be expected to lose an advantage progressively.

There was very limited genetic variation for traits specifically affecting growth under water stress. If most genotypes had a relatively lower rate of growth under water stress, the relative final yield of genotypes under water stress would be largely determined by differences in growth rates expressed when there was limited water stress. While some genetic variation in traits affecting growth under water stress may exist in sugarcane and related germ-plasm, large deviations in the expression of these traits may have occurred in none or very few genotypes.

5. Physiological mechanism of sugarcane during water stress condition

ABA, regulatory signaling molecules, implicated to stomata closure, reduction of leaf and stem growth, production of deeper root system, higher root and shoot hydraulic conductivity, assimilate remobilization, induction of senescence, maintenance of turgor pressure, expression of antioxidant proteins and seed dormancy [36]. There is concomitant relationship between ABA and stomatal conductance and transpiration rate, by increasing H_2O_2 raises cytosolic Ca^{2+}, concentration in guard cells [37] and by reducing ROS generation and the expression of antioxidant enzymes under stress conditions [38]. Additionally, ABA-mediated responses involve cis-acting elements as dehydration response and trans-acting factors as physiological and stress response regulators that induce ird29A gene, ScbZIP29 and ScbZIP31, ScbZIP21, ScbZIP24, ScbZIP70, and ScbZIP79 for controlling the gene expression [39] that regulates the water status in bundle sheath cells [40].

Reactive oxygen species (ROS), are key regulators for growth, development, response to biotic and abiotic stimuli and programmed cell death [41] as they are the byproducts of metabolic reaction of plants [42]. The exposure of the plant for drought for longer period of drought causes ROS outburst that overrides antioxidant mechanisms and damage cell membranes, DNA, and proteins and results in cell death [43] and low induce activate acclimation and defense pathways [44]. Plant antioxidant defense mechanisms are classified as enzymatic and non-enzymatic ROS scavenging. The enzymatic ROS scavenging consists of superoxide dismutase (SOD) activity and ascorbate peroxide (APX) and catalase (CAT) activities that may be genotype dependent, and display antioxidant response in sugarcane plants during stress conditions. On the other side, non-enzymatic oxidant molecules composed of ascorbic acid (AA), reduced glutathione (GSH), α-tocopherol, carotenoids, phenolics, flavonoids, and proline act as a compatible osmolyte, molecular chaperone and carbon and nitrogen reserve, and balance cytosolic pH. Increased proline is correlated to water tolerance as sugarcane acts as a component of antioxidant defense system rather than an osmoregulator.

The most oxidative stress symptom is peroxidation of lipids, which includes O_2 molecules that are originated from photosystem II and are incorporated into plasmid membranes and catalyzed by lipoxygenase (LOX) into LOOH (lipid hydroperoxide) causing membranes vulnerable to fragmentation and leading to a cascade of damaging events [45]. The product malondialdehyde (MDA) and thiobarbituric acid reactive substances (TBARS) cause changes in cell membrane properties, such as fluidity, ion transport, and enzyme activity [46] and are accepted as a marker of oxidative stress in plants [47]. A low level of lipid peroxidation revealed the water stress tolerance in sugarcane so it can be the best parameter to identify water stress tolerant sugarcane varieties.

6. Gene related to expression of water stress response in sugarcane

Several studies has been conducted to better understand the molecular basis of physiological responses of sugarcane under stress condition. Based on ABA-dependent, high throughput gene expression under water stress focused in extensive signal transduction networks that involve various transcription factors, protein kinases, and phosphates [48]. Expressed sequence tags (ESTs) are important and breakthrough tools for identifying sugarcane genes and assessing their function because of their complex genome size of 7440 Mb. Sugarcane Expressed Sequence Tags Sequencing project (SUCEST) sequenced over 238,000 ESTs from different sugarcane tissues and cultivars, and grouped them into 43,000 Sugarcane Assembled Sequences (SAS) [49] that provides important transcriptome studies [50]. The study conducted by Li et al. used cDNA array for studying the gene expression profile in sugarcane leaves subjected to water stress conditions. However, genes involved in cellular metabolism (cell wall, amino acid, lipid, and protein metabolism), signal transduction (transcription factors, hormone signaling proteins, calmodulins, and kinases), stress response (heat shock proteins and peroxidase) showed substantial similarity in expression under different experimental conditions, the expression pattern of genes varies on the intensity of water deficit condition.

The experiment to correlate the gene expression and drought tolerance of genotypes SP83–5073 (water stress tolerant) and SP90–1638 (sensitive) by use of macro-array 9 [51] from leaf library of SUCEST suggested that 93.3% expressed genes up-regulated in tolerant cultivars, where 36% of expressed gene were repressed un sensitive plants including heat shock proteins and genes involved in photosynthesis which is corroborated to morpho-physiological data. The interesting finding of this experiment opens a new research field to unravel the hitherto unknown genetic mechanism underpinning water stress tolerance in sugarcane. The fact of expression of antisense transcripts in sugarcane under water stress caused more complex in gene regulation under stress conditions [52] so, for instance, expression of miRNAs analysis done to cope with growth phases into different intensities of stress [26]. The gene expressed in field conditions provided a different set of genes and expression profiles as compared to glass-house conditions [53]. Since there are unclear remarks of drought stress gene and sucrose accumulation [10] caused the complexity in both phenomenon. Despite the sugarcane transcriptome response, [10] indicated the strong correlation between the expression of water stress-induced genes and the sequence of dehydrin (late embryogenesis abundant proteins for protection of cellular membranes and organelles during dehydration) so it can be used as molecular marker for water stress response in sugarcane [53]. On other hand, from root samples [54] detected gene coding proteins with protection function (chaperones, heat shock proteins, antioxidants enzymes, and protease inhibitors protein and ABA response (trehalose-phosphate synthase and serine/threonine kinase receptors) that are responsible for water stress protection and adaptation mechanism. Additionally, de Silva et al. verified water channel proteins, aquaporins with isoform PIP1-1, NIP3-1, and SIP1-2 in root were exclusively up-regulated in tolerant varieties for stress avoidance mechanisms in sugarcane [55–58].

Different genes encoding GAPDH, ά-tubulin and histone H1, eEF-1 ά, Eukaryotic initiation factor 4 ά, CUL (cullin), CAC (clathrin adaptor complex), APRT (adenine phosphoribosyltransferase) and TIPS-41 (Tonoplastic Intrinsic Protein 41) are good candidate genes for normalization in various stress caused by hormones, abiotic and drought. Among them, *CAC + APRT, GAPDH + eEF1*, and *CAC + CUL* are the most reliable genes for normalization of sugarcane under stress.

There is a lack of mutant or genetic lines available to confirm the gene expression that is identified by transcriptomic analysis. The translatability of glasshouse pot study to field is very little known so further reverse or forward genetic studies are the impetus for functional assessment and linking them.

7. Conclusion

Despite the numerous studies conducted on behalf of water stress conditions in sugarcane, the result was not significantly remarked. The response and expression shown by the sugarcane at a different intensity, location and genotypes varies and confused the exact reason and corroborate to another consequential sequence, it's all due to its complexity in genome structure. The protein dehydrin encoding gene was finally named as paradigm molecular marker for analysis of water stress in sugarcane. The morphological, physiological study cannot resolve the problem of water stress and developing tolerant varieties so the mutant line generation was necessary for comparing or studying the water stress tolerance. Besides, some *CAC + APRT, GAPDH + eEF1*, and *CAC + CUL* are providing reliable genes for normalization of sugarcane under stress. So further study in greenhouse and field conditions should be translated and the concept of reverse and forward genetics studies should be thoroughly implemented for developing stress-tolerant sugarcane varieties.

Author details

Abhisek Shrestha, Bharti Thapa* and Ganga Dulal
Agriculture and Forestry University, Bardibas, Nepal

*Address all correspondence to: thapa.bharti16@gmail.com

References

[1] FAOSTAT. FAO Statistical Database (FAOSTAT). Italy: Food and Agriculture Organization of the United Nations; 2014

[2] Inman-Bamber N, Smith D. Water relations in sugarcane and response to water deficits. Field Crops Research. 2005;**92**:185-202. DOI: 10.1016/j.fcr.2005.01.023

[3] Inman-Bamber NG, Lakshmanan P, Park S. Sugarcane for water-limited environments: Theoretical assessment of suitable traits. Field Crops Research. 2012;**134**:95-104

[4] Smit M, Singels A. The response of sugarcane canopy development to water stress. Field Crops Research. 2006;**98**: 91-97. DOI: 10.1016/j.fcr.2005.12.009

[5] Subramanian KS, Tenshia JSV, JayalakshmiK,RamachandranV.Antioxidant enzymeactivitiesminarbuscularmycorrhizal (*Glomusin traradices*) fungus inoculated and non-inoculated maize plants under zinc deficiency. Indian Journal Microbiology. 2011;**51**:37-43. DOI: 10.1007/s12088-011-0078-5

[6] Abdel-Fattah GM, Asrar AA. Arbuscular mycorrhizal fungal application to improve growth and tolerance of wheat (*Triticumaestivum* L.) plants grown in saline soil. Acta Physiology Plant. 2012;**34**:267-277. DOI: 10.1007/s11738-011-0825-6

[7] Borie F, Rubio R, Morales A. Arbuscular mycorrhizal fungi and soil aggregation. Revista de la Ciencia del Suelo y Nutrición Vegetal. 2008;**8**:9-18. DOI: 10.4067/S0718-27912008000200003

[8] Kaur K, Kaur N, Gupta AK, Singh I. Exploration of the antioxidative defense system to characterize chickpea genotypes showing differential response towards water deficit conditions. Plant Growth Regulation. 2013;**70**:49-60

[9] Chandra A, Dubey A. Effect of ploidy levels on the activities of Δ1-pyrroline-5-carboxylate synthetase, superoxide dismutase and peroxidase in *Cenchrus* species grown under water stress. Plant Physiology and Biochemistry. 2010;**48**:27-34

[10] Iskandar HM, Casu RE, Fletcher AT, Schmidt S, Xu J, Maclean DJ, et al. Identification of drought-response genes and a study of their expression during sucrose accumulation and water deficit in sugarcane culms. BMC Plant Biology. 2011;**11**:12. DOI: 10.1186/1471-2229-11-12

[11] Cutler SR, Rodriguez PL, Finkelstein RR, Abrams SR. Abscisic acid: Emergence of a core signaling network. Annual Review of Plant Biology. 2010;**61**:651-679

[12] Iqbal N, Nazar R, Khan MIR, Masood A, Khan NA. Role of gibberellins in regulation of source-sink relations under optimal and limiting environmental conditions. Current Science. 2011;**100**:998-1007

[13] de Silva A, Jifon JL, Da Silva JA, Sharm V. Use of physiological parameters as fast tools to screen for drought tolerance in sugarcane. Brazilian Journal of Plant Physiology. 2007;**19**:193-201

[14] Robertson MJ, Wood AW, Muchow RC. Growth of sugarcane under high input conditions in tropical Australia. I. Radiation use, biomass accumulation and partitioning. Field Crops Research. 1996b;**48**:11-25

[15] Carr MKV, Knox JW. The water relations and irrigation requirements of sugar cane (Saccharum officinarum): A review. Experimental Agriculture. 2011;**47**:1-25

[16] Du YC, Kawamitsu Y, Nose A, Hiyane S, Murayama S, Muraya S, et al. Effects of water stress on carbon exchange rate and activities of photosynthetic enzyme in leaves of sugarcane (Saccharum sp.). Functional Plant Biology. 1996;**23**:719-772

[17] Meinzer FC, Grantz DA. Stomatal and hydraulic conductance in growing sugarcane: Stomatal adjustment to water transport capacity. Plant, Cell and Environment. 1990;**13**:383-388

[18] Moore PH. Physiological basis for varietal improvement in sugarcane. In: Proceedings of the International Symposium on Sugarcane Varietal Improvement. Coimbatore, India: Sugarcane Breeding Institute; 1987. pp. 19-56

[19] Robertson MJ, Bonnett GD, Campbell JA. Effects of temperature on leaf area expansion in sugarcane: Controlled environment, field and model analyses. Field Crops Research. 1996a;**48**:23-29

[20] Saliendra NZ, Meinzer FC. Genotypic, developmental and drought-induced differences in root hydraulic conductance of contrasting sugarcane cultivars. Journal of Experimental Botany. 1992;**43**:1209-1217

[21] Kumar A, Sharma P. Climate change and sugarcane productivity in India: An econometric analysis. Journal of Social Development Science. 2014;**5**:111-122

[22] Chandiposha M. Potential impact of climate change in sugarcane and mitigation strategies in Zimbabwe. African Journal of Agricultural Research. 2013;**8**:2814-2818

[23] Thapa-Kshetri B. National seed legislation: Harmonisation with international treaties and conventions. Agronomy Journal of Nepal. 2013;**3**:138-149

[24] Lakshmanan P, Robinson N. Stress physiology: Abiotic stresses. In: Moore PH, Botha FC, editors. Sugarcane: Physiology, Biochemistry, and Functional Biology. Chichester: John Wiley & Sons, Inc.; 2014. pp. 411-434

[25] Kingston G. Benchmarking yield of sugarcane from estimates of crop water use. In: Proceedings of the Australian Society for Sugar Cane Technologists. Bundaberg; 1994. pp. 201-209

[26] Gentile A, Dias LI, Mattos RS, Ferreira TH, Menossi M. MicroRNAs and drought responses in sugarcane. Frontiers in Plant Science. 2015;**6**:58. DOI: 10.3389/fpls.2015.00058

[27] Moreira J, Goswami D, Zhao Y. Bioenergy-successes and barriers. In: Proceedings of Ises Solar World Congress 2007: Solar Energy and Human Settlement. Vol. I–V. Berlin, Heidelberg; 2007. pp. 38-45

[28] Walter A, Galdos M, Scarpare F, Seabra J, Leal M, Cunha M, et al. Brazilian sugarcane ethanol: Developments so far and challenges for the future. WENE. 2013;**3**:70-92. DOI: 10.1002/wene.87

[29] Inman-Bamber N. Sugarcane water stress criteria for irrigation and drying off. Field Crops Research. 2004;**89**:107-122. DOI: 10.1016/j.fcr.2004.01.018

[30] Delorge I, Janiak M, Carpentier S, Van Dijck P. Fine tuning of trehalose biosynthesis and hydrolysis as novel

tools for the generation of abiotic stress tolerant plants. Frontiers in Plant Science. 2014;**5**:147. DOI: 10.3389/fpls.2014.00147

[31] Daniels BA, Trappe JM. Factors affecting spore germination of the vesicular-arbuscular mycorrhizal fungus, Glomus epigaeus. Mycologia. 1980;**72**:457-471

[32] Bryla DR, Duniway JM. Effects of mycorrhizal infection on drought tolerance and recovery in safflower and wheat. Plant and Soil. 1997;**197**:95-103

[33] Wu QS, Zou YN, Wang GY. Arbuscular mycorrhizal fungi and acclimatization of micropropagated citrus. Communications in Soil Science and Plant Analysis. 2011;**42**:1825-1832

[34] Gemma JN, Koske RE, Roberts EM, et al. Mycorrhizal fungi improve drought resistance in creeping bentgrass. Journal of Turfgrass Science. 1997;**73**:15-29

[35] Comas LH, Becker SR, Von Mark VC, et al. Root traits contributing to plant productivity under drought. Frontiers in Plant Science. 2013;**4**:442

[36] Han C, Yang P. Studies on the molecular mechanisms of seed germination. Proteomics. 2015;**15**:1671-1679. DOI: 10.1002/pmic.201400375

[37] Pei Z-M, Murata Y, Benning G, Thomine S, Klüsener B, Allen GJ, et al. Calcium channels activated by hydrogen peroxide mediate abscisic acid signalling in guard cells. Nature. 2000;**406**:731-734. DOI: 10.1038/35021067

[38] Jiang M, Zhang J. Water stress-induced abscisic acid accumulation triggers the increased generation of reactive oxygen species and up-regulates the activities of antioxidant enzymes in maize leaves. Journal of Experimental Botany. 2002;**53**:2401-2410. DOI: 10.1093/jxb/erf090

[39] Schlögl PS, Nogueira FTS, Drummond R, Felix JM, De Rosa VE, Vicentini R, et al. Identification of new ABA-and MEJA-activated sugarcane bZIP genes by data mining in the SUCEST database. Plant Cell Reports. 2008;**27**:335-345. DOI: 10.1007/s00299-007-0468-7

[40] Sugiharto B, Sakakibara H, Sugiyama T. Differential expression of two genes for sucrose-phosphate synthase in sugarcane: Molecular cloning of the cDNAs and comparative analysis of gene expression. Plant & Cell Physiology. 1997;**38**:961-965. DOI: 10.1093/oxfordjournals.pcp.a029258

[41] Luis A. ROS and RNS in plant physiology: An overview. Journal of Experimental Botany. 2015;**66**:2827-2837. DOI: 10.1093/jxb/erv099

[42] Ahmad P, Jaleel CA, Salem MA, Nabi G, Sharma S. Roles of enzymatic and nonenzymatic antioxidants in plants during abiotic stress. Critical Reviews in Biotechnology. 2010;**30**:161-175. DOI: 10.3109/07388550903524243

[43] Miller G, Suzuki N, Ciftci-Yilmaz S, Mittler R. Reactive oxygen species homeostasis and signalling during drought and salinity stresses. Plant, Cell & Environment. 2010;**33**:453-467. DOI: 10.1111/j.1365-3040.2009.02041.x

[44] Mittler R, Vanderauwera S, Gollery M, Van Breusegem F. Reactive oxygen gene network of plants. Trends in Plant Science. 2004;**9**:490-498. DOI: 10.1016/j.tplants.2004.08.009

[45] Skorzynska-Polit E. Lipid peroxidation in plant cells, its physiological role and changes under

heavy metal stress. Acta Societatis Botanicorum Poloniae. 2007;**76**:49-54. DOI: 10.5586/asbp.2007.006

[46] Sharma P, Jha AB, Dubey RS, Pessarakli M. Reactive oxygen species, oxidative damage, and antioxidative defense mechanism in plants under stressful conditions. Journal of Botany. 2012;**2012**:26. DOI: 10.1155/2012/217037

[47] Pagariya MC, Devarumath RM, Kawar PG. Biochemical characterization and identification of differentially expressed candidate genes in salt stressed sugarcane. Plant Science. 2012;**184**:1-13. DOI: 10.1016/j.plantsci.2011.12.002

[48] Rabbani MA, Maruyama K, Abe H, Khan MA, Katsura K, Ito Y, et al. Monitoring expression profiles of rice genes under cold, drought, and high-salinity stresses and abscisic acid application using cDNA microarray and RNA gel-blot analyses. Plant Physiology. 2003;**133**:1755-1767. DOI: 10.1104/pp.103.025742

[49] Vettore AL, da Silva FR, Kemper EL, Souza GM, da Silva AM, Ferro MI, et al. Analysis and functional annotation of an expressed sequence tag collection for tropical crop sugarcane. Genome Research. 2003;**13**:2725-2735. DOI: 10.1101/gr.1532103

[50] Rodrigues FA, Da Graça JP, De Laia ML, Nhani-Jr A, Galbiati JA, Ferro MIT, et al. Sugarcane genes differentially expressed during water deficit. Biologia Plantarum. 2011;**55**:43-53. DOI: 10.1007/s10535-011-0006-x

[51] Rodrigues FA, de Laia ML, Zingaretti SM. Analysis of gene expression profiles under water stress in tolerant and sensitive sugarcane plants. Plant Science. 2009;**176**:286-302. DOI: 10.1016/j.plantsci.2008.11.007

[52] Lapidot M, Pilpel Y. Genome-wide natural antisense transcription: Coupling its regulation to its different regulatory mechanisms. EMBO Reports. 2006;**7**:1216-1222. DOI: 10.1038/sj.embor.7400857

[53] Ferreira TH, Gentile A, Vilela RD, Costa GG, Dias LI, Endres L, et al. microRNAs associated with drought response in the bioenergy crop sugarcane (*Saccharum* spp.). PLoS One. 2012;**7**:e46703. DOI: 10.1371/journal.pone.0046703

[54] Vantini JS, Dedemo GC, Jovino Gimenez DF, Fonseca LF, Tezza RI, Mutton MA, et al. Differential gene expression in drought-tolerant sugarcane roots. Genetics and Molecular Research. 2015;**14**:7196-7207. DOI: 10.4238/2015.June.29.13

[55] Da Silva ALC, and Da Costa WAJM. Varietal variation in growth, physiology and yield of sugarcane under two contrasting water regimes. Tropical Agricultural Research. 2004;**16**:1-12

[56] Basnet BMS. Rice: Water, food security and climate change in Nepal. Hydro Nepal: Journal of Water, Energy and Environment. 2012;**11**(1):78-80

[57] Bates LS, Waltren RB, Teare ID. Rapid determination of free proline for water – Stress studies. Plant and Soil; Australis. 1973;**39**:205-207. DOI: 10.1007/BF00018060

[58] Giovannetti M. and Mosse B. An evaluation of techniques for measuring vesicular arbuscular mycorrhizal infection in roots. New Phytologist. 1980;**84**(3):489-500